A Question and Answer Guide to As

Are we alone in the Universe?
Was there anything before the Big Bang?
Are there other universes?
What are sunspots?
What is a shooting star?
Was there ever life on Mars?

This book answers all these questions and hundreds more, making it a practical reference for anyone who ever wondered what is out there, where does it all come from, and how does it all work?

Written in non-technical language, the book summarizes current astronomical knowledge, without overlooking the important underlying scientific principles. Richly illustrated in full color, it gives simple but rigorous explanations.

Pierre-Yves Bely is an engineer specializing in the design and construction of large optical telescopes. He was Chief Engineer for the Canada-France-Hawaii Telescope, has worked on the Hubble Space Telescope and the design of its successor.

Carol Christian is an astrophysicist and Deputy of the Community Missions Office at the Space Telescope Science Institute. In addition to technical and outreach support of NASA missions, she is a collaborator on the Google Sky and World Wide Telescope projects for exploration of the sky on the Internet.

Jean-René Roy is an astrophysicist specializing in the evolution of galaxies and the formation of massive stars. He is Senior Scientist at the Gemini Observatory, which hosts two of the largest telescopes in the world, one in Hawaii and the other in Chile.

A Question and Answer Guide to
Astronomy

Pierre-Yves Bely
Carol Christian
Jean-René Roy

CAMBRIDGE
UNIVERSITY PRESS

CAMBRIDGE UNIVERSITY PRESS
Cambridge, New York, Melbourne, Madrid, Cape Town, Singapore,
São Paulo, Delhi, Dubai, Tokyo, Mexico City

Cambridge University Press
The Edinburgh Building, Cambridge CB2 8RU, UK

Published in the United States of America by Cambridge University Press, New York

www.cambridge.org
Information on this title: www.cambridge.org/9780521180665

First published 2008 as *250 réponses à ros questions sur l'astronomie* by La Compagnie des
Éditions de la Lesse.

English translation published 2010
Reprinted 2011

Printed in the United Kingdom at the University Press, Cambridge

A catalog record for this publication is available from the British Library

Library of Congress Cataloging-in-Publication Data

Bely, Pierre-Yves.
A question and answer guide to astronomy/Pierre-Yves Bely, Carol Christian, Jean-René Roy.
       p.   cm.
ISBN 978-0-521-18066-5 (pbk.)
1. Astronomy–Miscellanea. 1. Christian, Carol, 1950– ll. Roy, Jean-René. lll. Title.
QB52.B45 2010
520–dc22   2009046650

ISBN 978-0-521-18066-5 Paperback

Contents

The Earth 72

The Moon 95

Life in the Universe   173

History of astronomy   193

Telescopes 209

Amateur astronomy 250

Preface

Human beings are curious by nature and have marveled at the night sky ever since our *Homo sapiens* ancestors first gazed up into the heavens. What is "up there"? Why do stars shine? How did the Universe begin? Does life exist elsewhere? What is on the other side of the Moon?

Astronomy is one of the oldest sciences, but modern physics and technology, coupled with observations from space, have recently generated a stupendous wave of new knowledge. Most of our earliest questions about the nature of the Universe have now been answered, and many unexpected, intriguing new findings have been made, findings that invite us to be both humble and bold. And one needs not be a professional astronomer or physicist to understand them.

Our intent in writing this book has been to offer to the general reader a summary of current astronomical knowledge, generously illustrated and provided with rigorous but simple explanations, while avoiding mystifying professional jargon.

The 250 "windows" on astronomy in this book do not exhaust the topic, but we hope that they will pique the curiosity of our readers and stimulate them to explore further, by navigating on the World Wide Web or by consulting some of the many fine publications on astronomy, such as those suggested at the end of this book. Most important of all, we hope that they will find renewed wonder in the night sky!

April 2009

Acknowledgments

We would like to thank Sally Bely for much assistance in the final editing and Hélène Allard for sharpening key concepts for the general reader. We are also grateful to Nathalie Bely and Robert Macpherson for several illustrations and their many useful comments.

We would like also to thank Vince Higgs and Jonathan Ratcliffe of Cambridge University Press for their support and editorial assistance.

Units and numbers

We have used the metric system almost exclusively. Conversion factors for English equivalents can be found in the appendix.

In astronomy, distances, times, and temperatures are truly "astronomical numbers," in which the long strings of zeros are awkward and cumbersome. We have therefore often used scientific notation, in which numbers are expressed in powers of 10. The exponent of 10 is the number of places the decimal point must be shifted in order to express the number in its full form (left for negative exponents, right for positive exponents). For example, $2.5 \cdot 10^3$ is 2500, 10^6 is 1 followed by 6 zeros, or one million, and 10^{-6} is 0.000 001.

Notations

Numbers between square brackets (i.e. [3]) apply to the list of references at the end of the book.

References to related questions are noted by the letter Q followed by the number of the question. For example, (Q. 30) refers to question 30.

Young stars in the Small Magellanic Cloud. Credit: NASA/ESA.

Stars

1 Why do stars shine?

Just as a piece of iron glows red or white hot when heated in a forge, stars shine because they are hot, very hot: millions of degrees at the core and thousands of degrees at the surface. Early on, this was thought to be the result of combustion, that the stars were burning in the same way that coal burns, but if that were the case, they would have lifetimes of only a few thousand years, whereas most stars live for billions of years.

The formidable amount of energy necessary for such long lifetimes comes from two sources: gravity while the star is forming, then nuclear fusion during the rest of its life.

Stars are formed from interstellar clouds of dust and gases, mostly hydrogen, that become progressively concentrated. In the first stage of a star's life, the force of gravity pulls the cloud into a spherical shape (Q. 3). This contraction – think of it as a falling inward – releases energy, just as an object falling on our foot transmits energy to us that we perceive as pain and bruising. As the gas and dust heat up, they start to glow, emitting light weakly in the infrared. Eventually, as the temperature of the gas continues to rise, it begins emitting visible light. The cloud has now become a young star.

As the interior of the collapsing sphere grows hotter and denser, the gas molecules break up into atoms, then the atoms lose their electrons and become ions. At that point the gas has become an electrically charged hot plasma composed of an equal number of freely moving, positively charged ions and negatively charged electrons. Finally, the core of the sphere becomes so dense and hot (15 million K) that the hydrogen nuclei begin to collide and fuse into helium.

Stars being formed inside a cloud of gas and dust (NGC 604). Each red dot is a new star – about 200 are visible. Their light, rich in ultraviolet radiation, excites the atoms in the cloud of gas, making it glow. Credit: NASA/ESA.

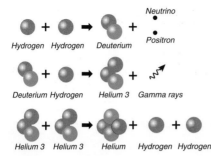

The three steps of hydrogen fusion into helium (the protons are shown in red, the neutrons in blue). Ultimately, four atoms of hydrogen have combined to form one atom of helium. The same result is also obtained via a chain of reactions with carbon, nitrogen, and oxygen acting as catalysts to convert hydrogen into helium.

Nuclear fusion[†] liberates enormous amounts of energy. Since the mass of a helium atom is 0.7% smaller than the mass of the four hydrogen atoms that formed it, a tiny amount of mass is "lost" for every helium atom produced. What happens to that mass? It is transformed into energy as per Einstein's famous equation, $E = mc^2$, which describes the equivalence of mass and energy. The mass that is transformed, m, may be minute but the speed of light, c, is very great (300 000 km/s), and its square is naturally much greater still. Thus, the product of the two terms, E, the equivalent energy, turns out to be enormous: the conversion of 1 kg of hydrogen into helium produces as much energy as burning 20 000 tons of coal. And the amount of hydrogen consumed in the stars is enormous, too: the Sun, for example, consumes 600 million tons of hydrogen every second! The total amount of energy produced is huge.

This energy produced in the core is propagated by radiation and convection towards the exterior layers of the star, and finally reaches the surface. The plasma at the surface then begins to radiate: the star shines.

The energy liberated in the interior of the star creates a pressure that combats and eventually counterbalances the force of gravity, so that the star ceases to collapse. At that point it stabilizes: it is an adult star.

The amount of energy transported to the surface – and therefore the star's temperature and color – is primarily dependent on the mass of the star (Q. 13). A star's lifetime also depends on its mass: the greater the mass, the shorter the lifetime. Large mass stars (10 or more solar masses) can sustain fusion for only a few million years; stars such as the Sun, for several billion years; small mass stars (7–10% of the mass of the Sun), for trillions of years. Objects of even lower masses cannot sustain fusion for very long and rapidly become warm cinders called brown dwarfs (Q. 50). These objects do glow in the infrared, however, due to energy released by their contraction.

2 What are stars made of?

Stars are huge balls of gas, primarily made up of hydrogen and helium. Hydrogen represents about 90% of the atoms in a star, helium slightly less than 10%. Both elements were produced in the Big Bang at the birth of the Universe, but nuclear fusion

---

[†] Nuclear fusion must not be confused with nuclear *fission*, in which a large atom, uranium for example, is split into two lighter atoms (Q. 32).

Schematic view of the interior of a star in which a helium core is forming. The core which is the size of Earth is actually much smaller than shown here.

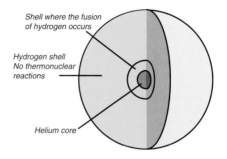

Shell where the fusion of hydrogen occurs

Hydrogen shell No thermonuclear reactions

Helium core

in stellar cores is constantly transforming hydrogen into more helium (Q. 1), changing the relative proportions of the two elements in stars over time.

The other elements found in stars, representing no more than 1% of the total, are oxygen and carbon, together with very small amounts of nitrogen, silicon, iron, copper, gold, silver, nickel, plutonium, and uranium. These elements may have been present in the original cloud out of which the star formed, or have been created later in its core.

Indeed, the very high temperature (from 10 million to several billion kelvins) and pressure in the interior of a star make it an alchemist's delight, where heavy elements such as carbon and oxygen and even silicon and iron are formed. If a core in which

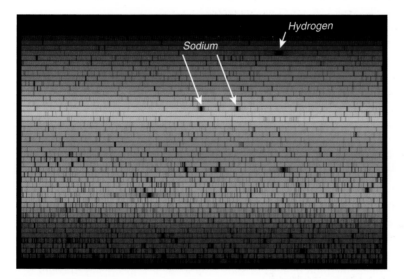

Spectrum of a typical star: the Sun. The bands, representing the wavelength ranges of visible light from purple to red, have been stacked on top of each other for compactness. Black areas are caused by the absorption by chemical elements present in the Sun's atmosphere, the darkness of their shade being a measure of their concentrations. One of the black areas in the red regions, for example, is due to the presence of hydrogen, the most abundant element in the atmosphere. The two black bands in the yellow region indicate the presence of sodium. Credit: Sharp, NOAO/NSO/Kitt Peak FTS/AURA/NSF.

hydrogen fusion is taking place reaches a high enough temperature, nuclear fusion based on helium begins. Three helium (He^4) atoms combine to form carbon (C^{12}), for example, then a carbon atom can combine with helium to form an atom of oxygen (O^{16}). The process continues, with the number of different reactions increasing over time. In some stars, the hydrogen fusion that began in the core eventually becomes a shell of hydrogen fusion moving outward through the body of the star, while helium fusion continues to take place in the interior.

We cannot see the composition of a star's interior, but we can use spectroscopy to determine the chemical elements at its surface[†] and also their relative percentages, because the different gases absorb light at very particular wavelengths.

3 Why are stars round?

Stars form when clouds of gas and dust coalesce under the influence of gravity. While the original cloud can be of any shape, the coalescing gas will eventually take on a spherical shape because any protuberance in the outer layers will exert pressure on the inner ones and tend to sink inward.

In reality, however, stars are not usually perfectly spherical because most stars rotate, and centrifugal force causes them to bulge out at the equator and be slightly flattened at the poles, as has happened with Earth.

4 How many stars are there in the Galaxy?

Our galaxy, the Milky Way (Q. 153), contains literally billions of stars – far too many to be counted one by one. Even the latest computer programs for automated deep digital sky surveys cannot count all the stars in our galaxy. We cannot even see all of them with our most powerful telescopes because some are obscured by gas and dust and some are very faint. Besides that, our solar system is embedded in our galaxy, and trying to determine the total number of stars from inside it is like trying to count the buildings in a large city by looking out of a window in a downtown apartment: we can see the buildings on the other side of the street and make out the upper stories of others further away, but our view of most of the buildings is blocked by other structures or veiled by haze in the distance. We do not even accurately know the number of stars

[†] Spectroscopy is the analysis of the wavelengths that make up the light from an object. This is similar to using a prism to spread light from a lamp into its constituent colors. The distribution of light intensity across a range of wavelengths is called a *spectrum*.

in our own solar neighborhood, which only extends out about 330 light-years, whereas the diameter of the Galaxy is of the order of 100 000 light-years.

Luckily, there are several ways to estimate the number of stars without having to count each one. One method involves first determining

Our solar neighborhood is the region that we can see well enough to analyze. It represents only about 5% of the Galaxy.

the overall brightness of our galaxy. Although things other than stars glow in the sky – luminescent gases and galaxies outside our own – their contribution to the brightness of the sky is small. We can then obtain the approximate number of stars by simply dividing the total brightness of the Galaxy by the average brightness of a star. How do we obtain the brightness of the Galaxy?

Our solar system is located about halfway along the radius of the Galaxy, meaning that we actually see a good part of the light emitted by the whole of it. Now, one of the best all sky images that we have comes from the Two Micron All Sky Survey (2MASS) in the infrared (Q. 154). Infrared light is similar to visible light but is lower in energy and usually associated with thermal emission (heat). One of the great advantages of infrared light is that it penetrates dust, "sees" further, and so provides us with a more complete picture. The value for the total brightness of the Galaxy derived from these infrared images is comparable to that for other galaxies similar to ours, confirming the validity of that measurement. As for the average luminosity of stars, we can obtain it by measuring the luminosity of stars in our own solar neighborhood whose distances we can evaluate, and thence derive their intrinsic luminosities (Q. 5).

A second estimate of the number of stars can be made by determining the mass of the Milky Way and dividing it by the mass of an average star. The mass of a galaxy can be determined from the influence of gravity on the gas, dust, and stars it contains. All the celestial bodies that make up the Galaxy rotate around the galactic center; and just as Newton's law of gravitational attraction allows us to calculate the movement of one body around another if we know the mass of the central body, we can determine the mass of the central body if we know the speed of rotation of a body in orbit around it.[†]

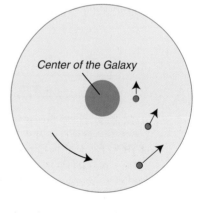

By this method we arrive at the total mass of the Galaxy, not only stars, but also interstellar dust and the invisible "dark matter" (Q. 148), so the result is an upper limit for the total mass of all stars.

[†] The force of gravitational attraction between two bodies of mass M and m separated by distance r is: $F = GMm/r^2$, where G is the gravitational constant. If the body of mass m revolves around the (larger) mass M, the force is $F = mv^2/r$. This "centrifugal" force is balanced by the force of gravity. Combining the two equations, we find that $M = v^2r/G$. Thus, if we know the velocity, v, of mass m, we can determine M.

Both methods provide approximately the same result, namely that our galaxy contains about 100 billion stars (10^{11}).

5 How are the luminosities of stars measured?

Astronomy is a very old science, and in its earliest days the brightness of stars was estimated subjectively with the naked eye. It was the second century BC Greek astronomer Hipparchus who, as he was compiling the first star catalog in history (Q. 192),[†] devised the scale known as "magnitudes" to categorize stellar brightness. He established six categories, in which the stars in each category appear to be twice as bright as those in the previous one. The brightest stars were assigned a magnitude of 1, while magnitude 6 stars were barely visible to the naked eye.

With the advent of photographic plates and more recently, of electronic devices like those in digital cameras, we can now assess brightness objectively. For the sake of continuity, the magnitude scale has been retained, but it is now defined in physical instead of physiological terms. As it turns out, human perception of auditory and visual stimuli follows an approximately logarithmic law. For example, if we hear a series of sounds whose intensities actually vary as the progression 1, 2, 4, 8, 16, our brain interprets it as intensities progressing by 1, 2, 3, 4, 5. We judge the last sound to be five times louder when it is actually sixteen times louder.[‡] For this reason the decibel scale for sound intensity follows a logarithmic law.

Since the same type of perception also applies to the eye's reaction to light intensity, the magnitude system in use today defines the magnitude of a star as being proportional to the logarithm of its intensity, with the proportionality constant slightly adjusted so that the magnitude of visible stars roughly corresponds to the naked eye classification of Hipparchus.[§]

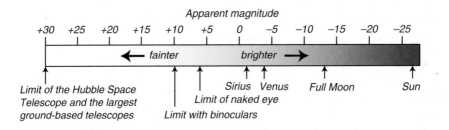

[†] An impressive feat for its time, his catalog contained relatively precise positions and apparent brightnesses for 850 stars.

[‡] This sensory peculiarity evolved in man, and in mammals in general, because it is advantageous: senses can collect a much wider range of intensities if the response is logarithmic than if the perception is linear.

[§] In this system, two objects with apparent fluxes ϕ_1 and ϕ_2 measured in the same conditions (i.e. in the same wavelength), have the magnitudes m_1 and m_2 such that: $m_1 - m_2 = 2.5 \log \frac{\phi_2}{\phi_1}$. A difference of five magnitudes corresponds to a brightness ratio of 100, and a difference of 10 magnitudes to a brightness ratio of 10 000.

The above diagram provides some guidance. Note that Sirius, the brightest star in the sky, does not have a magnitude of 1, as per Hipparchus, but now has a negative magnitude (−1.4). The faintest objects we can observe with current large optical telescopes have a magnitude of 30, i.e. are 1000 billion times less luminous than the brightest stars. A telescope 10 m in diameter receives only a dozen photons per second from such objects and it requires hours of exposure to detect them accurately.

The magnitude system is counterintuitive because the fainter the star, the larger the magnitude number. Besides that, the magnitude of a star actually depends on the specific wavelength and bandwidth, i.e. the range of wavelengths, observed. So currently, the tendency is to use the more intuitive scale used in radio astronomy, the jansky, which is a measure of the energy received from a star (in watts) per unit of surface area (square meter) and of frequency observed (hertz).[†]

6 How are the distances to stars measured?

The nearest stars are so far away that we cannot hope to measure their distances by using radar, as we do for the Moon, but we can use the method that surveyors employ to determine the distance to a remote hilltop or church steeple. They measure the change in the direction of their landmark when viewed from two points separated by a known distance, called the base (Q. 79). With the two angles and the base length known, the triangle is completely determined and the distance to the object can be calculated. For stars, where the distances involved are so great, as long a base as possible must be used in order to obtain sufficient precision. And the longest base at our disposal is the diameter of the Earth's orbit.

The position of the target star is therefore observed relative to much more distant background stars at a certain time of year. Over the following six months, the Earth completes half an orbit around the Sun, creating a base line of approximately 300 million km. The star is then

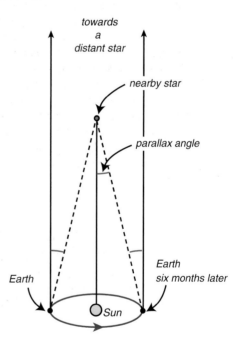

observed from this new vantage point, again relative to the background stars, and its apparent shift in position allows the determination of the star's distance as described above. The angle subtended by the radius of the Earth's orbit as seen from the star is called the *parallax*, and this stellar triangulation method bears the same name. If the parallax is 1 arcsecond (e.g. 1/3600 of a degree), the distance to the star is 3.26 light-years ($3 \cdot 10^{13}$ km) which is called a *parsec* (Q. 7).

[†] A jansky, abbreviated Jy, is equal to 10^{-26} W m^{-2} Hz^{-1}.

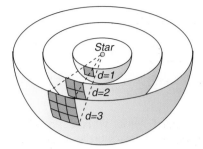

A source of light such as a star illuminates an area that grows as the square of the distance, so the apparent luminosity of the star decreases as the inverse of the square of that distance.

This method works for distances up to about 500 light-years. Beyond that, the parallax is too small to be measured with current instruments. The best measures have been obtained from space by the Hipparcos satellite, which was able to determine the distance of stars up to 650 light-years away with an accuracy of 5%.

For more distant stars, we must turn to indirect methods, the most common being estimating a star's distance from its apparent brightness. Stars come in many types and colors, but it turns out that, in general, stars of the same color shine with the same intrinsic luminosity (Q. 14).[†] Since the apparent brightness of a light source decreases as the square of its distance, the distance of a star can be calculated from a comparison of its apparent brightness to its intrinsic brightness as estimated for its type. For example, if we find a star similar to the Sun, we can estimate its distance from its apparent brightness, since we know the intrinsic brightness of the Sun.

This method works best with a very special class of stars called "standard candles," whose intrinsic luminosity can be determined with great accuracy from characteristics other than color. A common example is the Cepheid variable, a class of stars whose intrinsic luminosities are related to their pulsation periods (Q. 18).

Unfortunately, measurements of stellar distances based on apparent brightness are affected by interstellar gas and dust that absorb some of the stars' light, thus making them appear dimmer than they actually are.

7 Parsecs? Light-years? Why not miles or kilometers?

The distance to the Moon, our closest neighbor in space, is 384 000 km, and to Alpha Centauri, the binary star nearest us, it is 41 500 000 000 000 km. Our minds can easily

---

[†] Astronomers make a subtle distinction between luminosity and brightness. The *luminosity* of a star is a measure of how bright it really is, while *apparent brightness* or just *brightness* is a measure of how bright it appears to us on Earth. More precisely, the luminosity of a star is the total amount of energy at all wavelengths and in all directions that it radiates per unit of time, and is expressed in watts, while apparent brightness is the amount of energy received per unit time and unit area and is expressed in W/m^2. To avoid possible confusion in this text, we generally qualify luminosity as *intrinsic luminosity*. Astronomers often express intrinsic luminosity in terms of magnitude, using the concept of *absolute magnitude*, which is by convention, the apparent magnitude a star would have if it were at a distance of 10 parsecs.

grasp the approximate magnitude of the first distance,
but for the second, the long
string of zeros baffles compre-
hension. We need something
more compact and intuitive.
We could always write such

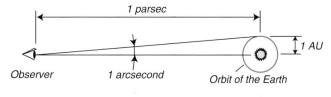

large numbers in scientific notation: $4.15 \cdot 10^{13}$ km, which is certainly more convenient
for making calculations, but is still not very easy to grasp. This is why astronomers have
adopted special units to describe cosmic distances.

Inside the solar system, the most convenient unit to use is the astronomical unit (AU),
which is defined as the average distance from Earth to the Sun (Q. 43) and is about
150 million km. This unit works well in our home system: Neptune, for example, the
most outlying planet, has an orbital diameter of 30 AU, and comets, in their vast orbits,
travel to a maximum distance of 100 000 AU from the Sun.

For stars and galaxies, the most commonly used unit is the parsec (pc) and its
multiples (kiloparsec and megaparsec). The parsec, which is an abbreviation for the
words *parallax* and *arcsecond*, is the distance at which an object would be located if it
had a parallax angle of 1 arcsecond, using a base distance equal to the radius of the
Earth's orbit.

The advantage of using the parsec is that it is intrinsically connected to the method
of measuring distances by parallax (Q. 6). If the parallax of a star is 1 arcsecond, its
distance is 1 parsec (1 pc $= 3.1 \cdot 10^{13}$ km). If the parallax is 10 times as small, or 0.1
arcsecond, the distance is 10 times greater, or 10 pc. For very great distances inside
the Galaxy, we measure in thousands of parsecs, or kiloparsecs (kpc), and for distances
to other galaxies, in millions of parsecs, or megaparsecs (Mpc). For example, the Sun is
8 kpc from the center of the Galaxy, and the Virgo cluster, the cluster of galaxies nearest
us, is at 15 to 20 Mpc.

Another common unit is the light-year (LY), the distance traveled by light in one year
(1 pc = 3.26 LY). It is a unit of distance, not of time as its name would suggest. It has
the advantage of incorporating information on the age of distant objects. For example,
if a galaxy is 1 billion LY away, we know that we are seeing the Galaxy as it was 1 billion
years ago. The limit of the visible Universe is 13.7 billion LY (Q. 134). If we detected an
object at this outer limit today, we would be seeing it as it was 13.7 billion years ago,
i.e. almost at the birth of the Universe.

8 How are the masses of stars determined?

The light from a star brings us a great deal of information. From it we can infer the
star's surface temperature, its diameter and chemical composition, but not its mass.
The only way to determine the mass of a celestial body is to observe the effect of its
gravitational pull on another object revolving around it. For example, observing the
dance of satellites in orbit around a planet allows us to determine the planet's mass.

Since we cannot determine the mass of a star in isolation, it is fortunate for us that
more than half of all stars are binaries. However, for a binary system to be useful to
us, it must be a "visual binary" (Q. 17), that is to say that we must be able to *visually*

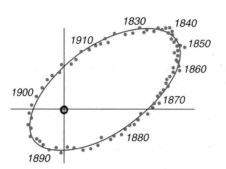

The orbit of the companion star of 70 Ophiuchi, with a period of 88 years.

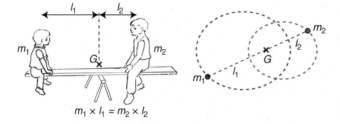

$m_1 \times l_1 = m_2 \times l_2$

distinguish the movements of both stars in order to map the orbit of one around the other.

The position of the stars in the sky are measured in terms of angles, so we must also determine the distance between Earth and the binary system in order to convert these angular measurements into actual distances between the two stars. This done, Kepler's third law then provides the sum of the two masses in the system (Q. 194).[†]

How do we obtain the mass of each of the two stars after determining the total mass of the system? In reality, one star does not revolve around the other; the two stars orbit around their common center of gravity (Q. 107). Their positions relative to the common center of gravity allows us to determine the ratio of their masses. For an analogy, imagine two children sitting on a seesaw. If they wish to stay in equilibrium, the heavier child must sit closer to the pivot. The distance of each child from the balance point is inversely proportional to the child's weight. Once we know the ratio of the masses of two stars and the sum of their masses, we can easily deduce the individual masses.

Unlike stellar diameters and luminosities which extend over a very wide range (one million and 10 billion times, respectively), the range of star masses is nowhere near that great, extending between about 1/15th and 150 times one solar mass. But even at the bottom of the scale a star has a lot of mass: the lightest stars are nearly 30 000 times more massive than Earth. True stars cannot exist with less than about 0.08 times the mass of the Sun because their gravity is insufficient to trigger nuclear fusion (Q. 1).

[†] Using the Sun/Earth system as reference, this law can be expressed as $m_1 + m_2 = a^3/P^2$ where m_1 and m_2 are the masses of two stars in Sun-mass units, a is the semi-major axis of the ellipse traveled by one of the stars around the other, expressed in astronomical units (the distance from Earth to the Sun – see Q. 43), and P is the period of revolution of the star in its orbit expressed in years.

9 How big are the stars?

Stars come in a great range of sizes. The smallest ones, neutron stars (Q. 15), are only a few tens of kilometers in diameter, while the largest supergiants have diameters hundreds of millions of kilometers across, 1 000 times the diameter of the Sun. If we exclude these exceptional cases and just consider normal stars in the main sequence (Q. 14), we find that they have diameters of between 1/10th and 10 times the diameter of the Sun.

Certain stars may appear much larger than others in photographs of the sky, but that does not mean they are really bigger. The effect is caused by the diffraction of light in the telescope (Q. 211), and overexposure which causes the brightest stars to have larger images.

Star diameters are actually extremely difficult to measure directly because normal telescopes do not have sufficient resolution to resolve stellar disks. Only a few nearby giants have been measured by the Hubble Space Telescope and by interferometers from the ground. Some diameters have also been determined thanks to an *occultation*, either by the Moon or by a companion star in a binary system. Although it is hard to measure the diameters of stars directly, it is relatively easy to calculate them indirectly using the laws of radiation. For a body radiating with a continuous spectrum, which is a good first approximation for most stars, the energy that is emitted per second per

The largest and smallest stars in the main sequence compared to the Sun. Red supergiants, which are 1000 times larger than the Sun, and white dwarfs, which are 100 times smaller, would be impossible to represent in the scale of this figure.

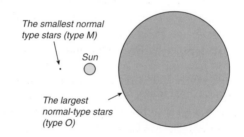

The smallest normal type stars (type M)

Sun

The largest normal-type stars (type O)

Image of the giant star *Mira A*, or Omicron Ceti taken by the Hubble telescope. It is a binary star, and material is being pulled into its companion. Credit: NASA/ESA.

Image of a portion of the sky containing both bright and faint stars. None of the disks are actually resolved, but the bright stars appear larger due to the diffraction of light by the telescope and to overexposure. Diffraction is also responsible for the cross-shaped spikes in the images. They are due to light being scattered by the thin blades supporting the telescope's secondary mirror (Hubble telescope in this case). Credit: NASA/ESA.

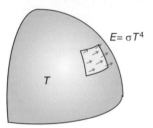

unit area is only a function of temperature (to the fourth power – according to Stefan's Law[†]). Therefore, if we know the intrinsic luminosity of a star, for example because we have measured its apparent luminosity and we know its distance to Earth, and further, if we obtain the star's temperature from its color (Q. 13), we can deduce its total surface area and thus its diameter.

10 How big do stars get?

The largest known star, the supergiant μ Cephei, 4900 LY from Earth, has a diameter almost 1500 times that of the Sun. If this monster occupied the Sun's position, it would encompass the orbit of Jupiter! The star Betelgeuse in the constellation of Orion, at a distance of 427 LY, is not far behind with a diameter of about 1000 times that of the Sun.

The most massive star ever found in our galaxy – and also the most luminous – is the Pistol star, which is responsible for the nebula of the same name. Its mass is now about 100 times that of the Sun, but it has already shrunk a great deal by shedding its outer envelope. At its formation 3 million years ago, it was about 200 times the mass of the Sun. The Pistol star is intrinsically 10 million times brighter than the Sun, and if it were not completely embedded in a thick cloud of dust, we would be able to see it with the naked eye as a fourth magnitude star. As it is, we can only see it in the infrared, which is able to penetrate dust clouds. The star's surface temperature is estimated at 100 000 K. Massive stars like these burn their hydrogen very quickly and have short lives (the Pistol

[†] If L, R, and T are, respectively, the absolute luminosity, radius, and temperature of the star under study, and if $L\odot$, $R\odot$, and $T\odot$ are those same parameters for the Sun, the Stefan–Boltzmann law can be expressed as $L/L\odot = (R/R\odot)^2(T/T\odot)^4$ from which the star radius R can be derived.

The Pistol star, the white dot in the center of the image, is the most massive and the brightest star yet discovered in our galaxy. It created the cloud engulfing it by twice ejecting its envelope. The cloud is so large (four light-years) that it would extend from our Sun to the next nearest star. This picture, taken in infrared light, is shown in false colors. Credit: Figer, UCLA/NASA.

consumes the same amount of hydrogen in one second as the Sun does in a year). The Pistol should explode as a supernova in about three million years.

11 How old are the stars?

The stars that we see today are not all the same age. The oldest are nearly as ancient as the Universe, about 13 billion years old, while others are still in the process of formation. Our own Sun is a "middle-aged" star approximately 4.5 billion years old.

The age of a star can be determined if we know its mass, temperature, and luminosity, with the relationship between these three quantities being well established from theoretical models and confirmed by observation. A star's evolution depends directly on its mass (Q. 14): the more massive it is, the faster it burns up its hydrogen. Once a star's mass is known, its luminosity and temperature enable us to determine where it is in its evolutionary "lifetime," and from that information we can estimate its age.

12 How old is the oldest star?

The oldest star ever discovered in our galaxy is called HE 1523-0901. It is slightly less massive than the Sun.

Such low-mass stars evolve very slowly (Q. 14). The age of HE 1523-0901 was calculated from traces of radioactive elements in its atmosphere, uranium and thorium in particular, which can be used to determine the age of celestial bodies just as carbon-14 can be used to date organic compounds on Earth. Calculations show that the star appears to have formed 13.2 billion years ago, only 500 million years after the Big Bang [20]. The small amount of uranium and other heavy elements in its makeup indicate that it does not belong to the first generation of stars, but inherited these elements from the explosion of a previous supernova (Q. 135). This type of old fossil star is very rare.

13 Do stars really come in different colors?

Yes, for people with very good eyes. Betelgeuse, in the constellation of Orion, is red. Rigel, in the same constellation, is blue, as are Sirius and Vega. The Sun itself is white, a neutral color – although, when close to the horizon, it appears yellow due to atmospheric absorption (Q. 121). The eye loses its sensitivity to color in low light – at night, for example – so that faint bright stars appear white to us, but in reality they are colored.

When a piece of iron is heated, it first turns red, then yellow, and finally white as its temperature increases. Similarly, the color of a star depends on its temperature – but the temperature at its surface and not at its core, which is much hotter (Q. 1). If its surface temperature is under 4000 K,[†] the star emits mostly in the infrared and appears reddish. If the temperature is over 7000 K, the star emits primarily in the ultraviolet and appears blue.

Detailed spectroscopic analysis of the light from stars, i.e. the measurement of its intensity as a function of wavelength, can tell us not only about surface temperatures, but also about chemical compositions.

Stars are categorized into seven main spectral classes that are identified by the letters O, B, A, F, G, K, M, in order of decreasing temperature.[‡] The disconcerting,

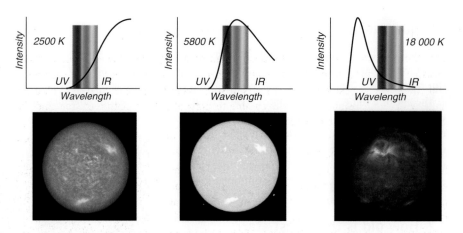

Stars emit simultaneously in all the colors of the rainbow, as well as in the infrared and ultraviolet, which our eyes cannot detect. Our perception of color in a star depends on the wavelength of its strongest emissions, and that, in turn, depends upon the temperature at the star's surface.

[†] The kelvin (symbol: K) is the unit increment of temperature of the kelvin scale (formerly called a degree kelvin) measured above the absolute zero, which is about $-273°C$. Absolute zero is the coldest temperature that matter can attain – at this temperature, the atoms stop vibrating altogether. $T(K) = t(°C) + 273.16$.

[‡] The classic mnemonic for remembering this series is: "Oh Be A Fine Girl, Kiss Me!" – the series has recently been augmented with the L, T, and Y classes for cooler dwarf stars.

Annie Jump Cannon (1863–1941) who, with a number of other women astronomers at the observatory at Harvard, spent nearly a lifetime analyzing hundreds of thousands of stellar spectra to determine their spectral classes. At the time, it was not considered appropriate for a woman to spend long nights outside observing through a telescope, and women astronomers were relegated to inside positions, in the laboratory.

unalphabetical sequence of letters derives from the original classification, which was based on the appearance of hydrogen, carbon, calcium and iron absorption lines in stellar spectra. At the time, the letters A through O ran in alphabetical order, but as measurements and interpretations were refined, giving us a deeper understanding of the relationship between the spectra and temperature, types O and B had to be moved in front of A, and some of the other classes were either eliminated or merged, leaving us with the jumbled-looking sequence we have today.

14 How many different kinds of star are there?

Over the last two centuries, careful observation of the sky has allowed astronomers to measure or estimate the brightness, mass, diameter, color (or more accurately, spectrum), and chemical compositions of thousands of stars. What have we learned from all these data? What conclusions have we been able to draw about the nature of stars and the principles governing their evolution?

Whenever one is faced with a massive amount of undigested data, the best way to tease out the under-lying relationships is to plot one type of data against another in a graph. For stars, the most telling graph, probably the most important of all modern astronomy, is the Hertzsprung–Russell diagram (abbreviated H–R). Developed independently in the years 1906–13 by Ejnar Hertzsprung, a Danish amateur astronomer who later became a professional, and the American astronomer Henry Norris Russell, this chart plots the intrinsic bright-ness of stars as a function of their temperature.

E. Hertzsprung and H. Russell

What is amazing is that 90% of all stars in the Universe fall into almost total alignment along a relatively narrow band in the diagram. Since it does cover the vast majority of observed stars, this band is called the *main sequence*, and in that band, the stars are perfectly ordered according to their masses; the cooler stars have low masses and the very hot stars are the most massive. Why should this be?

It really turns out to be no mystery, once we realize that the most fundamen-tal property of a star is, precisely, its *mass*. A star's mass governs all its other

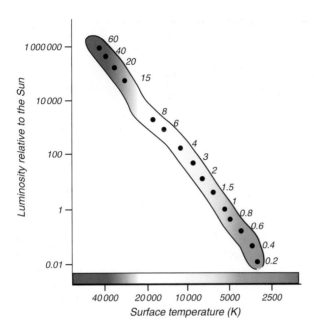

The Hertzsprung–Russell diagram for main sequence stars, in which luminosity is plotted against surface temperature. The numbers along the main sequence line refer to stellar masses, expressed in solar masses. The diagram can also be drawn up using the x-axis to show spectral classes or the color index of stars, each variant slightly changing the shape of the main sequence line.

characteristics: surface temperature, intrinsic luminosity, diameter, and lifetime. This is true because stars shine thanks to the energy released during their conversion of hydrogen into helium (Q. 1); the greater the mass, the greater the effect of gravity, and the greater the gravity, the higher the pressure and temperature in the core of the star. Hence, the more intense the thermonuclear fusion process, which in turn raises the surface temperature, hence the luminosity, of the star. Note that stellar diameters also increase with mass, as they do with temperature and luminosity (Q. 9).

Massive stars consume their hydrogen more quickly, which means that their lifetimes are short. The most massive stars (type *O*) typically only live for three to four million years, whereas stars of lower mass (type *M*) can survive for thousands of billions of years, i.e. hundreds of times the current age of the Universe! Small stars (with a mass less than 0.8 times that of the Sun) are by far the most numerous and represent approximately 90% of all the stars in the main sequence. Very massive stars (over eight times the mass of the Sun) are rare.

Some stars are not even on the main sequence, however. Just as with human populations, where some individuals are strikingly taller, shorter, or heavier than the average, a small percentage of stars falls well outside stellar norms: there are dwarfs, giants, even supergiants. Most of the stars that are off the main sequence are white dwarfs (9% of the total number), while only 1% are giants and supergiants. Stars are not born as dwarfs or giants, but become that way with age as they exhaust their reserve of nuclear fuel (Q. 15). On the extreme right of the diagram are stars that are just being formed, called *protostars*, that will eventually move onto the main sequence.

The distribution of the different types of star in the Hertzsprung–Russell diagram. The majority (90%) are in the main sequence. Stars outside it have either left the sequence at the end of their lives (large green arrows) or are just being born and are moving towards (and will eventually join) the sequence (small green arrows at right).

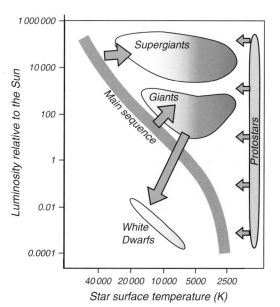

15 How do stars die?

A star dies when it has used up all of its nuclear fuel, but the details of its demise will vary, depending on its initial mass.

In stars of very low mass (less than 0.4 times one solar mass), the heat of the core diffuses mainly by convection, i.e. by the movement of gases that are heated by coming into contact with the core and, as a result, becoming less dense, rising up to the surface where they cool down, then plunging once again towards the core to repeat the cycle. These movements continuously feed hydrogen into the core, sustaining the nuclear fusion taking place there, and making the core grow bit by bit. But once all the hydrogen is consumed, the star has been transformed into a ball of inert helium gas. This process is extremely slow, taking hundreds of billions of years.

In stars of intermediate mass like our Sun, the energy from the core flows by radiation in the inner region, and convection is restricted to the outer layer, meaning that only the hydrogen in the core can be consumed. Once that hydrogen is exhausted, the core contracts under pressure from upper layers. This in turn raises the core's temperature and allows hydrogen fusion to proceed in a spherical shell surrounding it. This new release of energy now forces the outer layers of the stellar atmosphere to expand, and the star has become a red giant. The process then repeats itself, this time with the outbreak of helium fusion in the core to produce carbon and oxygen. Simultaneously, there is further expansion of the outer layers, making the star even larger. When the core's helium is exhausted, it contracts. This heats it up again, and it now expels its outer atmosphere to

Small mass star

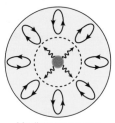

Medium mass star

The planetary nebula, Abell 39, with at its center the star that ejected it. The nebula is an envelope of hydrogen and other trace elements illuminated by the light of the star. Credit: NOAO/AURA/NSF.

form a *planetary nebula.*[†] Eventually, all that remains of the original is a small star with a hot, dense core of carbon and oxygen, called a *white dwarf.*

Finally, when the mass of the star is large, i.e. at least eight times as large as the Sun, it, too, enters a giant phase once its hydrogen core is consumed, but in this case the event is unimaginably violent and leaves the star a *supergiant.* The temperature in the core of a supergiant is higher than in stars of average mass, and nuclear fusion extends beyond the synthesis of carbon. Eventually, the star self-destructs in a supernova explosion (Q. 16). The core then implodes and the residue is an extremely dense object that can be either a neutron star or a black hole.

A neutron star is a sphere about 10 to 20 km in diameter in which the density of matter is so high that the protons and electrons have merged to form neutrons. Without electrical charges to repel each other, the neutrons can then be compacted to the point of being in actual contact with one another, giving the object phenomenal density. If the Earth were compressed to such an extent, it would fit nicely inside a football field.

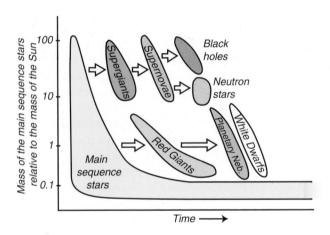

Schematic view of the evolution of stars according to their mass. The timescale is relative.

[†] Despite the name, planetary nebulae have nothing to do with planets. The term originated in the nineteenth century at a time when observations through small telescopes suggested that these objects were large gassy planets like Jupiter.

And as for black holes, here we are dealing with the ultimate concentration of matter (Q. 145).

Low-mass stars can live for hundreds of billions of years. Stars of average mass, like the Sun, are stable and spend about 10 billion years on the main sequence, then 100 million years in the red-giant phase. High-mass stars spend about 70 million years on the main sequence, then 5 million in their supergiant phase. Supernova explosions only last for about 10 s, but the residual object remains bright for months.

16 What is a nova? A supernova?

A nova – the name means "new" in Latin – is a star that suddenly becomes enormously bright. Novae were so named because they appeared where no star had been seen before, but that was simply because they had been too faint to be visible to the naked eye. And when they did become visible, it was because they had undergone a violent nuclear explosion.

Most novae are the result of an explosion in a binary star system (Q. 17) in which one member of the pair has already exhausted its hydrogen to become a white dwarf (Q. 15), and the other is a normal "main sequence" star that has exhausted the nuclear fuel in its interior, its outer layers have expanded, and it has become a red giant. As the atmosphere of the aging giant expands, the material is captured by its dwarf companion. Such a transfer of matter onto a stellar surface is called accretion.

The explosion occurs as the material from the red giant is deposited on the surface of the white dwarf. Compressed under the white dwarf's gravity and heated to the point of triggering nuclear fusion, the accreted material releases a vast amount of energy as it explodes, blowing the gases away from the white dwarf at incredible speeds, up to thousands of km/s, and causing a sudden brightening of the binary pair by a factor of 50 000 to 100 000. Although the brightening is dramatic, only a small amount of the total mass of the system is ejected – about 1/10 000th of the mass of the Sun. The process is sometimes repeated periodically: the binary star RS Ophiuchi, for example, has exploded approximately every 20 years over the past century. Its last explosion occurred in 2006.

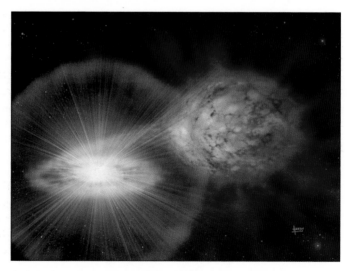

Artist's view of a nova event. The white dwarf on the left is accumulating matter from its companion, a red giant. The accreted matter explodes in a nuclear fusion reaction when it reaches the surface of the white dwarf. Credit: D. Hardy.

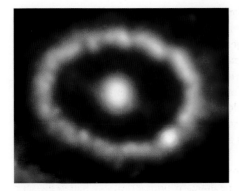

The supernova SN 1987A appeared in February 1987 in the Large Magellanic Cloud, a companion galaxy to our own. The event actually occurred 160 000 years ago, but the light has taken that long to reach us. The glowing ring is produced by the shock wave from the supernova as it encounters gas left behind by previous events. The image was obtained with the Hubble Space Telescope in 1994. Credit: C. Burrows, NASA/ESA/STScI.

If the white dwarf accretes a very large amount of material from its companion, it undergoes "runaway" nuclear fusion, an event that completely destroys the star in an explosion even more gigantic than that produced in a mere nova: this is a supernova.

Such an explosion in which a star is completely destroyed can also occur at the end of the life of a single massive star. For a star of between one and eight solar masses, life ends with it ejecting most of its material in the form of a planetary nebula (Q. 15), leaving only the core. Very massive stars, those over eight solar masses, annihilate even their cores in their death throes. Here is how it happens.

In a star of modest mass, when the hydrogen fuel in its core has been exhausted, energy continues to be produced by the fusion of helium into carbon and oxygen, but the reactions stop there. In truly massive stars, the pressure and temperature become so intense that the fusion continues, at first with oxygen being merged to form silicon, and then, in continued fusion events, with the production of elements up to iron.

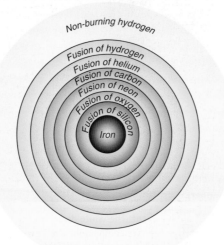

Diagram of layers of fusion in the core of an older, very massive star. The extremely dense nucleus, approximately the size of Earth, is tiny compared to the size of the star.

Fusion processes must stop once the stellar core has become iron because the merging of elements heavier than iron actually consumes energy rather than releases it (Q. 32).

At this point, since energy is no longer being produced in the core, the internal pressure drops and gravity forces cause the outer layers to collapse. The massive star suddenly implodes, compressing the core to the point where its protons and neutrons are squeezed into close contact with each other. The density of such a core is enormous: a teaspoon of this degenerate matter would weigh 400 million tons. The core responds with an explosion of incredible violence, sending a titanic shock wave throughout the star. The explosion can be so stupendous and the star's collapse so complete that the result can be the creation of a black hole.

The explosion blows off a significant amount of hot material which expands rapidly outward at 5000–20 000 km/s, producing the dramatic brightening of a supernova. The brightening can be five billion times the brightness of the Sun.

The matter ejected during the explosion is so hot that many nuclear reactions are triggered and a series of heavy chemical elements are produced. In fact, supernovae are the primary source of heavy elements in the Universe, including plutonium, uranium, and other exotic elements. This material, rapidly ejected into the surrounding space, eventually drifts into contact with clouds of gas and dust in interstellar space and can eventually be incorporated into new stars and planetary systems like our own.

The most spectacular supernova in our galaxy in historical times occurred in 1054. Noted by Chinese and Korean observers, the bright new star lasted several weeks and was visible even during the day. The next supernova in our galaxy could very well be Eta Carinae, but nobody can predict when this might happen. It could be in the next few years – or in a million years. The mass of this star is 100 times that of the Sun, and it has already begun to manifest large variations in brightness.

Of the several types of supernova that are recognized, depending upon the exact mechanism that produces the explosion, the Type Ia is of particular interest. The time it

The Crab Nebula, the remains of the supernova reported by the Chinese in 1054 AD. Its diameter is enormous, 11 LY, and it is expanding at 1500 km/s. In the center is a pulsar (Q. 19), a neutron star that rotates at 30 times per second and emits strong gamma radiation. Credit: NASA/ESA.

Eta Carinae, a massive star in the sky of the southern hemisphere. The two lobes, which are the size of our solar system, are composed of gas ejected in an explosion that occurred 150 years ago. Credit: Morse/NASA/ESA.

takes for supernovae to brighten rapidly and then dim has a profile called a *light curve*. As it happens, the maximum intrinsic luminosity of Type Ia supernovae is remarkably uniform, with some subtle adjustments depending upon how steeply the light curve declines. This allows astronomers to use them as "standard candles" to measure distances. If such a supernova is discovered in a distant galaxy, its intrinsic luminosity can then be determined which, when compared to its observed brightness, yields the distance to that galaxy. This is the same method as that used with Cepheid variables (Q. 18), but supernovae are much brighter, and can reach deeper into the Universe.

17 What is a double star?

Double stars are stellar bodies that appear to form a pair in space. But there are doubles and doubles ... Some of these pairs are false doubles, not actually adjacent to each other, just located in the same direction in the sky as seen from Earth. True (or intrinsic) double stars are also called *binary stars*, and these are physically associated. They orbit around each other, drawn together by the force of gravity, and, in most cases, they even formed together.

We do not know precisely the proportion of binary stars in the sky, but it may be

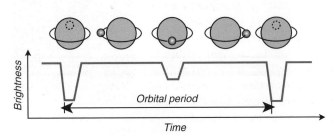

as high as two out of every three stars, and triple, quadruple, and even quintuple star systems also exist. The distance between the two stars can be very small, with their atmospheres almost in contact, or as large as several thousand astronomical units.

Binaries are discovered in a variety of ways. Sometimes both stars can be visually distinguished – those are called *visual binaries*. Other binary systems are detected by spectroscopy: they may be so close visually that they appear to be a single object, but when their combined spectrum is observed repeatedly, their individual motions can be detected due to the Doppler effect (Q. 141) – these are called *spectroscopic binaries*. Other such systems are discovered because their brightness varies periodically with time, as one star passes in front of the other. These are the *eclipsing binaries*.

Binary stars are of vital importance in astronomy because they provide our only chance to measure the masses of stars using Kepler's laws of orbital physics (Q. 8).

18 What are the Cepheids?

Most stars shine with almost constant brightness – the Sun's brightness, for example, varies by only 0.1% over a period of 11 years. Yet some stars vary significantly in brightness and over much shorter time periods. Some of them shrink and expand as their internal structure changes because they are exhausting the fuel in their interior, a type of variation called *pulsation*.

One of the most common and best studied types of pulsating star is the *Cepheid variable*. These are older stars, usually yellow giants, that vary quite regularly over a period of 1 to 50 days. They brighten and dim as they physically expand and contract, acting like a spring. The gas of the star's envelope compresses due to gravity, causing it to heat and brighten. Eventually pressure overcomes gravity, the envelope of gas rebounds, and the temperature drops causing the brightness to decrease. Then, the pressure decreases, gravity takes over, and the cycle begins again. One might expect that the pulsations would eventually stop because of energy losses in each cycle, but in fact, a complex phenomenon due to the ionization of helium in the star's envelope sustains the pulsation.

The first variable star of this type was discovered in 1784 by John Goodrick, a Dutch amateur astronomer who paid for the discovery with his life: he caught pneumonia while making his long nocturnal observations – he was not yet 22 when he died. The object of his interest was the star δ Cephei in the constellation Cepheus, which gave its name to this class of variable star.

Cepheids play a crucial role in astronomy because, as demonstrated by the American astronomer Henrietta Swan Leavitt in 1912, the period of a Cepheid's variation is related to the intrinsic luminosity of the star, and can thus serve as a "standard candle" to measure large distances (Q. 6). Cepheids are very bright, with up to 10 000 times the luminosity of the Sun, making it possible to detect them even in relatively distant galaxies. Once a Cepheid is found and monitored, its pulsation period supplies its intrinsic luminosity. The comparison of the intrinsic luminosity to the apparent brightness provides the distance to the star's home galaxy. It was this method that Edwin Hubble used in 1929 to demonstrate the expansion of the Universe (Q. 130), and was again used recently with the Hubble Space Telescope in order to determine its rate of expansion with greater precision.

Henrietta Swan Leavitt (1868–1921).

19 What is a pulsar?

A pulsar is a stellar object that appears to be emitting intermittently at radio wavelengths. In reality, it is a rapidly spinning neutron star emitting two constant beams of radio waves that we perceive as pulses as they sweep past the Earth – just as we perceive the sweeping beam of a lighthouse as discrete flashes of light.

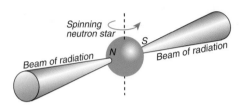

The first pulsar was discovered in 1967, and more than 1500 of them are now known. Their period is typically about 1 s, but some rotate as fast as once per millisecond, 100 times faster than a power drill. The period can be extremely stable, constant to within a few seconds over a million years – more accurate than the best atomic clocks on Earth.

A pulsar forms when a supernova explodes, leaving a neutron star in place of the original stellar core (Q. 16). The original star revolved slowly on its axis (like our Sun, which rotates once in 27 days), but as it collapsed into an extremely small, compact neutron star, its speed increased – like an ice skater pirouetting (Q. 162) – resulting in the incredibly fast rotational speeds found in pulsars.

The two radio beams issue from "hot spots" on the surface of the neutron star that are associated with the magnetic poles. They are the result of the phenomenon of *synchrotron radiation*, thus called because it was first noticed in a synchrotron (subatomic particle accelerator). This kind of electromagnetic radiation occurs when electrons spiral through magnetic fields at speeds close to the speed of light.

20 Do stars ever collide?

The globular cluster M80 contains several hundred thousand stars, and collisions probably do occur. Credit: NASA/ESA.

In our neighborhood, the stars are so far apart that the risk of collision is very low. In globular clusters, however, where thousands of stars are crowded into a relatively small volume of space, collisions may be frequent.

Evidence that this is actually happening in star clusters is provided by the presence of abnormally blue stars in them. When two stars do merge, they form a single, massive, very hot star which can be recognized by its intense blue color. All of the stars in a globular cluster were born at the same time, and any blue ones born then would already have "died" because massive stars such as these have very short lifetimes. Therefore, the presence of such a star, called a "blue straggler," in a globular cluster could only be explained by its being a youthful new star born of a collision.

Simulation of the collision of two old stars, transforming them into a blue straggler. Credit: Brown/Lombardi.

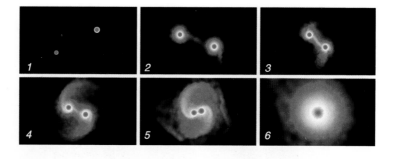

21 Are we really made of stardust?

When the late American astronomer Carl Sagan said "We are made of stardust," he was not just waxing poetic. The chemical elements of which we are made, really were produced inside of ancient stars billions of years ago. To make us even more humble, these elements – except for the hydrogen – are in fact only the ashes, residues, debris, of the violent processes that light up the stars.

During the Big Bang, the only elements created were essentially hydrogen, helium, and a tiny amount of lithium. When the first generation of stars formed, the other chemical elements were produced through normal fusion and, later, through the violently explosive processes that occur in supernovae (Q. 16). The elements blown out of those dying stars eventually mingled

Carl Sagan (1934–96).

with the hydrogen and helium that make up the gas clouds of the interstellar medium. The carbon and silicon bonded with oxygen and nitrogen to form small particles, like dust, of silicates and other compounds. Under the tug of gravity, this mixture of gas and dust aggregated to form new stars. The process goes on still, with stars forming, living, dying, scattering their ashes, then forming anew. Most stars in our galaxy – including the Sun – belong to at least the third generation of star formation.

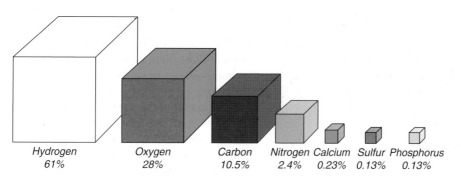

| Hydrogen | Oxygen | Carbon | Nitrogen | Calcium | Sulfur | Phosphorus |
| 61% | 28% | 10.5% | 2.4% | 0.23% | 0.13% | 0.13% |

The main chemical constituents of human beings (in number of atoms). Hydrogen was synthesized in the Big Bang; all other chemical elements were produced inside stars.

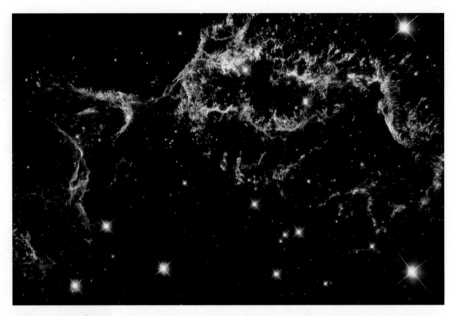

Remains of the titanic dismemberment of a star: a supernova. This is Cassiopea A which exploded 10 000 years ago, but whose light reached us only in 1680. We recognize the various chemical elements by the color they emit: dark blue regions are rich in oxygen, red, rich in sulfur, and the white, pink, and orange regions contain mixtures of oxygen and sulfur. All these chemical elements will be recycled in the next generation of stars, perhaps incorporated into new planets and ... living beings? Credit: NASA/STScI.

During the birth of our own solar system, most of the aggregated material in the primordial cloud collapsed into the nascent Sun, but some matter escaped that fate and was left to form a swirling disk of gas and dust in orbit around the new star. The planets, including Earth, then coalesced out of this disk material. The elements in that original cloud were the raw materials for our oceans, our land masses, and eventually all forms of life on our planet. Without the stars, the material we are made of would not exist; nor would we!

22 Do all civilizations recognize the same constellations?

Today, we have no problem knowing what day and month it is. We have calendars, watches, newspapers, television, and computers to keep us informed. But what about 4000 years ago? Back then, there was only the sky ...

Knowing one's way around the sky was very useful in the past; it was an excellent calendar, if one knew how to read it. It could tell the farmer when to plant, the herder when to move his animals to new pastures, the shaman when to repeat his rituals. To distill sense from the myriad bits and bytes of information contained in the starry firmament, nothing was easier for pre-technical peoples than to pick out patterns – persisting, recurring patterns – in the stars. Looking up at the sky on an August night in the northern hemisphere, we can see the summer patterns: Lyra, Cygnus, the swan,

The Big Dipper is a configuration of stars which is quite obvious. This pattern has been used by a number of civilizations, but with different interpretations.

and Aquila, the eagle. In December, those patterns are absent, but we know that they will come back, and that when they do, summer will be back, too.

We call these starry patterns "constellations," from the Latin *cum*, meaning with, together, and *stella*, star. The ancient Sumerians gave us the constellations Taurus (in Latin, the bull), Leo (the lion), and Scorpius (the scorpion), animals that were important in their culture. The Greeks linked stellar patterns with their myths: Orion, the hunter with his dogs, and the Pleiades, the "Seven Sisters." When European navigators discovered the sky of the southern hemisphere, they saw "the Telescope," "the Microscope," "the Clock." And if it was up to us to baptize a new pattern in the sky today, we might "see" a car, a plane, or ... Elvis.

Not all peoples saw the same patterns in the stars. Amerindians, Mayas, African tribes, the ancient Greeks, the Chinese, all imagined different images using different groupings of heavenly bodies. The Big Dipper is an exception. The pattern it makes in the sky of the northern hemisphere is so distinctive that many different peoples have recognized it – not that they saw the same object in it. For the Greeks and certain Amerindians, the pattern evoked a bear. For the ancient Chinese, it was the carriage of the emperor of the celestial world. In medieval Europe, it was a horse-drawn cart. For Americans today, it is a dipper, and for the British, a plough.

Of course, the stars in a given constellation are not physically linked and are distributed in three-dimensional space, so if there are any other civilizations in our general neighborhood of the Galaxy, and if they should search for shapes in their sky, they would see different patterns even if they looked at the same stars as we do. And they would be unlikely to use the same stars for their constellations, but would certainly pick out different ones that would be brighter for them. In matters like this, perspective is everything.

23 How many constellations are there?

Even though the constellations no longer play the same important role in modern societies as they did in ancient ones, they remain a convenient way to divide the sky into different areas. Astronomers of the world have therefore agreed to partition the sky into 88 mutually agreed-upon constellations. These official constellations include

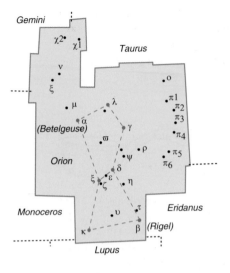

The limits of the constellation Orion (light blue) on the sky. The stars in the figure of Orion himself are represented in blue, with the other main stars in black. As is customary, the brightest stars are identified by a Greek letter (Q. 24).

most of those in the northern hemisphere known to the ancient Greeks. As a whole, they are an interesting mix: 14 men and women, 9 birds, 2 insects, 19 land animals, 10 marine animals, a centaur, a unicorn, a dragon, a winged horse, a river, and 29 inanimate objects, including a furnace, a compass, and a pump.

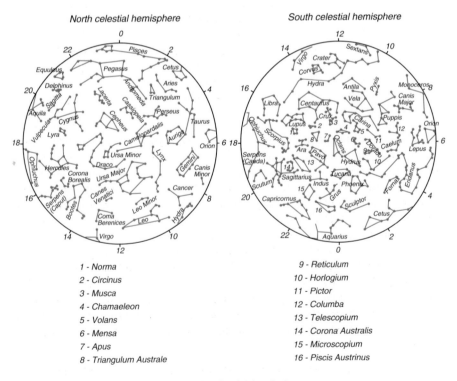

1 - Norma	9 - Reticulum
2 - Circinus	10 - Horlogium
3 - Musca	11 - Pictor
4 - Chamaeleon	12 - Columba
5 - Volans	13 - Telescopium
6 - Mensa	14 - Corona Australis
7 - Apus	15 - Microscopium
8 - Triangulum Australe	16 - Piscis Austrinus

The constellations of the north and south celestial hemispheres.

In order to cover the whole sky, the constellations have been assigned specific limits of meridians and parallels (right ascension and declination) to surround the symbolic central figure, resulting in a patchwork effect, like a set of puzzle pieces fitted together. In scientific parlance, the constellations are referred to by their Latin name. For example, the Big Dipper is referred to as *Ursa Major*.

24 How are stars named?

In the western world, the brightest stars had all been given names by ancient Greek times. After the fall of the Roman Empire, however, astronomy was only practiced as a science by the Arabs. They rescued the body of Greek astronomical science, summarized in Ptolemy's *Almagest* (Q. 192),[†] from potential oblivion, and adopted the Greek constellations, but they assigned their own names to many of the brightest stars. When texts were later translated from Arabic into Latin in the later Middle Ages, some of those names became garbled or were changed. Other cultures had their own names for stars, of course, but for European society, the Arabic names were retained when Europeans resumed their pursuit of the astronomical sciences during the Renaissance. The most brilliant stars thus have Arabic names, and many of those have poetic meanings. The following table gives the common name and translation of some Arabic names for stars, together with their scientific names.

Name	Arabic name	Meaning	Bayer designation
Aldebaran	Ad-Dabaran	Follower (of the Pleiades)	α Taurus
Altair	At-Ta'ir	The Eagle	α Aquila
Betelgeuse	Yad al-Jauza'	The hand of Orion	α Orionis
Deneb	Dhanab ad-Dajajah	The tail of the Swan	α Cygnus
Eltanin	At-Tinnin	The Dragon	γ Draconus
Rigel	Ar-Rijl	The foot of Orion	β Orionis

It would be an exercise in futility, of course, to try to give a colloquial name to every star in the sky; professional astronomers had to invent coding systems to designate stellar objects. In the 1600s, Johann Bayer, a German lawyer from Bavaria, invented a system in which stars in the different constellations were labeled in order of descending brightness, but using the Greek symbols and letter order in the alphabet. For example, the star Alpha (designated by the Greek letter α) Centauri, our closest stellar neighbor, is the brightest star in the constellation Centaurus (Q. 29).

For fainter stars, it is difficult to determine by eye which are brighter than others. By 1725, John Flamsteed, Astronomer Royal of England, had decided to establish his catalog of more than 3000 stars by simply assigning a number to each star in a constellation, for example, 61 Cygni or 47 Ursae Majoris. This system has been widely adopted, but the brightest stars are still referred to using the Bayer system or even by their Arabic names. So a bright star may have several names, in addition to catalog numbers.

[†] The Arabs were so impressed by Ptolemy's work that they called his book *Al Magister*, the Grand, which later became Almagest.

When astronomers began to use large telescopes, ever fainter stars could be observed, and catalogs were compiled by various institutions. Since these catalogs overlap, some stars have a great many names! The most common catalogs in use are the *Bonner Durchmusterung* from the Bonn Observatory, published in 1859, the *Smithsonian Astrophysical Observatory Catalog*, and the *Henry Draper Catalog*, established in the 1920s. Stars are referred to by their number in the catalog, preceded by the prefixes *BD, SAO,* and *HD,* respectively. The *Henry Draper Catalog* is particularly widely used because it contains more than 200 000 stars along with their spectral classifications.

Today, the most precise way to refer to a star is by giving its coordinates in the sky, but since coordinates change with time, owing to precession of the equinoxes (Q. 87), a reference date for the coordinate system must be added, typically 1950 or 2000. And since stars move on their own as well, the actual year of the observation must also be indicated.

25 Can we still discover and name stars?

The International Astronomical Union (IAU) coordinates the work of astronomers around the world and is the only organization authorized to name celestial objects. Newly discovered objects (stars or otherwise) are named by their coordinates and date of discovery.

Commercial offers to let you put a name of your choice on a star (or other celestial object) are all, without exception, completely fraudulent – even those offering to provide you with a "baptismal certificate."

26 Is there a southern polar star?

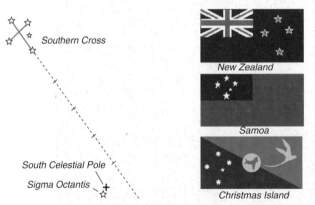

There is indeed a star at the south celestial pole, Sigma (σ) Octantis, also called the southern polar star, but it is barely visible to the naked eye and thus is not a good marker for navigation. On the other hand, the southern sky contains the Southern Cross, the most prominent feature of the heavens "down under," and the long axis of the Southern Cross points roughly to σ Octantis and so can help locate the South Pole. The Southern Cross must not be confused with the *false cross* nearby. The true Southern Cross contains a fifth star that is missing in the false cross.

The south celestial pole is in line with the major axis of the Southern Cross, about 4.5 axis lengths south of the constellation. The beautiful Southern Cross has been incorporated into many national flags, including those of Australia and New Zealand, and the flags of several islands in the Pacific Ocean.

The Southern Cross is visible throughout the year in the southern hemisphere for observers south of 34° south latitude. It can also be seen from the lower latitudes of the northern hemisphere in the early part of the night in April and May, including from North Africa and the Caribbean. The Ancient Greeks knew about this constellation because, in their day, it was visible from Athens. This is no longer the case because of the precession of the equinoxes (Q. 87).

27 How many stars are visible to the naked eye?

The total number of stars visible to the naked eye from Earth is estimated to be approximately 8000, half of which are visible from the northern hemisphere and half from the southern. But at any one time, in either hemisphere, only about 2000 stars can be seen because the other 2000 are then in the daytime sky. To even see the night-sky 2000 requires a very dark site, however, well away from any city, and the eye must be fully dark adapted, which means a wait of 20 to 30 min in complete darkness.

Observing visually with a small amateur telescope or binoculars, will show many more stars, especially in the Milky Way. But stars will remain points of light and do not come more detailed. The most interesting aspect in observing through amateur telescopes is that we can observe the planets and some beautiful nebulae that cannot be seen easily with the naked eye (Q. 239).

A panorama of the sky showing most of the stars visible to the naked eye. This is actually a drawing in which 7000 individual stars are shown as white dots, with size indicating brightness. The Large and Small Magellanic Clouds are the two fuzzy patches in the lower right quadrant. Credit: Lund Observatory.

28 Are the stars fixed or do they move?

Ancient astronomers believed that the stars were fixed in the sky because that is what they observed. From one night to the next, one year to the next, the stars always seemed to occupy the same positions relative to each other. This distinguished them from the Sun, Moon, and planets, which very definitively moved relative to the field of background stars. The word planet actually means "wanderer" in Greek.

In fact, stars are not stationary: they are in orbit around the center of the Galaxy, as is the Sun. Their orbits are all slightly different from the Sun's, however, and so they appear to move relative to us. In astronomy, the apparent slow drifting of stars across the sky is called their *proper motion*.

The stars in our solar neighborhood are speeding along at a lively clip, in the order of 200 km/s, but since the Sun is moving, too, and the stars are very far away, their apparent speeds are very low, in the order of 0.1 arcsec/yr – a very small motion to measure. The record for relative speed is held by Barnard's Star, with a proper motion of 10 arcsec/yr.

Most of the stars in our familiar constellations are not physically associated,[†] are relatively far from each other, and therefore orbit around the center of the Galaxy at different speeds. As a result, the appearance of our constellations change slowly over time. The figure shows how the Big Dipper will look in 50 000 years:

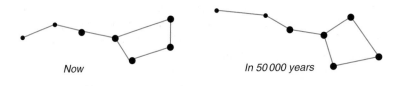

Now   *In 50 000 years*

29 Which star is closest to us?

Our solar system's nearest neighbor is Proxima Centauri, which is part of the three-star Alpha Centauri system located in the southern hemisphere in the constellation Centaurus, near the Southern Cross. Alpha Centauri is about 4.3 LY or 40 trillion km from the Sun. The two brightest stars in that system are so close together that we perceive them as a single bright object. They are in orbit around each other at a distance of 23 AU (23 times the distance between the Sun and Earth – note that Uranus is 19 AU from the Sun). These two stars complete an orbit around each other every 79 years. They are both similar to the Sun and are named Alpha Centauri A and Alpha Centauri B. The third star in the system, Proxima Centauri, is also known as Alpha Centauri C because it is the third in brightness. It is a type *M* red dwarf star that may be in orbit around the other two, although it is quite far (0.21 LY) from the pair, and it is not even

---

[†] There are exceptions such as the Pleiades, also known as Messier 45 or the "Seven Sisters." Those stars are part of a "star cluster," i.e. a collection of stars that formed together, are relatively close to each other, and all move at the same speed.

absolutely certain that Proxima Centauri is bound by the force of gravity to the other two stars in the system. It is so faint that it was not discovered until 1915.

Proxima Centauri is now the closest star to the Sun and has been so for about the last 32 000 years. It will remain the closest for another 33 000 years, but as it is moving away from us relatively quickly, it will then be replaced by another star, Ross 248.

30 Between stars that die and stars that are born, is the population of our galaxy growing or shrinking?

The current world population is a little under 7 billion human souls. At 100 billion, the number of stars in our galaxy might appear quite enormous (Q. 4) relative to that. But while the number of human births is approximately 130 million per year, the rate of star formation in the Galaxy is much lower by comparison: the equivalent of about four solar masses, or about seven new stars per year (with more small stars than medium or large ones being born) [15]. This rate is much lower than it used to be because the amount of hydrogen available for star formation decreases as our galaxy ages: since its birth, 10 billion years ago, our galaxy has converted about 90% of its primordial gas into new stars.

At what rate are the stars dying? The massive stars end their lives in supernovae, and they do so at the rate of about one star every 50 to 100 years in our galaxy.[†] The less massive stars such as the Sun, which are much more numerous, go nova and end their lives as planetary nebulae and white dwarfs at a rate of about one per year.

There you have it: if one star, more or less, dies each year and seven are born, the population of the Galaxy is growing, albeit slowly.

31 Are there any isolated stars, outside of the galaxies?

Imagine a night sky with no stars, blank, black – with at most a small cloudy spot or two, all that can be seen of our closest neighbor galaxies. Sad, no? This is the fate of any who may live on planets that orbit around lonely stars, drifting through the vast reaches of empty, intergalactic space.

When two galaxies collide, millions of stars are created from the froth born of billions of stars getting tossed wildly about, as the stellar landscape is completely and violently transformed. But when the mayhem subsides, most stars end up with companions and neighbors. A small percentage, however, may be torn from their parent galaxies and set adrift in intergalactic space. Several hundred such orphan stars have been detected in the Virgo cluster of galaxies about 60 million LY from here, and there must be many others that share the same fate.

[†] But across the whole observable Universe, it is estimated that a supernova explodes every second.

Between these two galaxies, M81 and M82, which are themselves embedded in a cluster of galaxies, a mighty gravitational battle has been raging for billions of years, doubtless leaving some lonely stars adrift in the intergalactic region between them. Credit: J. Schedler.

32 Could nuclear fusion, the process that fuels the stars, be tamed to solve our energy problems?

Most of the radiation from stars is produced when two hydrogen atoms fuse to form a helium nucleus (Q. 1). Huge amounts of energy are released in this process. Why not use it to produce power on Earth?

Our nuclear power plants currently produce electricity through nuclear *fission*. What is the difference between fission and fusion? Nuclear fission occurs when a large atom, uranium for example, is split into two lighter atoms, while in nuclear fusion two lightweight atomic nuclei merge to form a heavier nucleus. How can fusion and fission, two processes which work in reverse ways, both produce energy?

In both cases the energy released is derived from the energy that binds protons and neutrons together inside an atomic nucleus. Protons and neutrons are kept together inside the nucleus by the "strong nuclear force" which overpowers the repulsive electrostatic force between positively charged protons and keeps protons and neutrons packed closely together. Of the four fundamental forces, gravity, electromagnetic force, weak force (responsible for radioactivity), and strong nuclear force, the strong nuclear force is the most powerful. However, it acts only over very short distances.

When a nucleus is formed from its individual *nucleons* (protons and neutrons), the mass of the resulting nucleus is less than the sum of the masses of the nucleons that make it up, a mass loss known as the "mass defect" (Q. 1). The energy released in assembling the nucleus from constituent nucleons is called the "nuclear binding

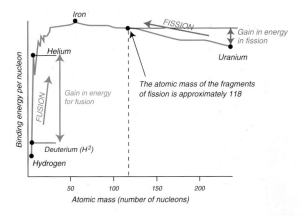

energy." It is the energy corresponding to the mass loss as per Einstein's mass/energy equivalence formula ($E = mc^2$).

The average binding energy per nucleon first increases and then decreases as larger and larger nuclei are formed. It is maximum for iron (Q. 16). Remembering that the binding energy is energy that is released in forming the nucleus, we can gain energy by transforming nuclei in such a way as to "climb" the binding energy curve. This can be done in two ways,

• by assembling nuclei lighter than iron into heavier ones – this is fusion;
• by breaking nuclei heavier than iron into lighter ones – this is fission.

For the production of energy on Earth, nuclear fusion of hydrogen would have significant advantages over the fission of heavy nuclei such as uranium or plutonium. First, as the curve shows, fusion produces far more energy.[†] Second, hydrogen (or deuterium), the fuel for fusion, is available in seawater in virtually limitless quantities. This is not the case with the precious element, uranium, required for fission. Finally, contrary to fission, fusion does not result in radioactive waste.

In order to make fusion occur, the mutual repulsion of nuclei must be overcome, and this requires very high temperatures (approximately 15 million K) and very high density of matter. These conditions are created in the H-bomb, but the chain reaction that occurs is uncontrolled. In civilian power implementations, the whole difficulty resides in controlling the conditions of temperature and density required. This situation is called "confinement." Two methods are currently being studied: inertial laser confinement and magnetic confinement. The latter was adopted by the ITER (International Thermonuclear Experimental Reactor) project. The technologies involved are very complex and industrial production is not expected to be reached for at least another 30 years.

[†] Even if, in practice, it appears preferable to use deuterium instead of hydrogen: although the energy production is lower, deuterium has an extra neutron so that the atom is bigger and therefore an easier target to bombard.

The Solar System

33 How did the Solar System form?

Immanuel Kant (1724–1804) and Pierre Simon de Laplace (1749–1827) independently proposed the hypothesis of the solar nebula to explain the formation of the Solar System. It is still considered the most plausible approach.

No one was there to witness the birth of our solar system, but close study of the different bodies that compose it – our Moon, other planets, asteroids, meteorites, and comets – together with observations of planetary systems in formation elsewhere in the Galaxy provide many clues.

It all began about 5 billion years ago with an interstellar cloud of gas and dust about 1 LY in diameter. This "solar nebula" was composed of primordial hydrogen mixed with residue from a supernova explosion. Under the action of gravity, the cloud began to contract, accelerated perhaps by a shock wave from another nearby supernova. Had the nebula been perfectly spherical and homogeneous, it would have contracted into a non-rotating sphere of gas. As it was, irregularities of shape and internal turbulence caused it to start rotating (figure (a)). As the cloud contracted further, its rotational speed increased, like a skater accelerating her pirouette (Q. 162). As the nebula became denser, its temperature increased due to collisions between its fast-moving molecules and atoms. Soon the young Sun had formed at the cloud's center (Q. 1), while the periphery became flattened under centrifugal force and took the shape of an extended disk (b and c).

When the young Sun was producing enough light and heat, radiation pressure put a stop to the contraction of the disk (Q. 46). The gas and dust in it cooled and separated. Collisions between small particles produced larger ones, then pebble-sized bodies, and eventually "planetesimals" kilometers in diameter.[†] Planetesimals with masses sufficient to attract each other eventually merged into protoplanets, and during the last phase of this process, which must have been spectacular, bombarded each other like so many balls on a busy billiard table, resulting in the formation of the interior planets, including Earth.

Many planetesimals did not merge into planets or satellites, however; they became Kuiper Belt Objects, comets, asteroids, and other bits of debris (Q. 69).

[†] This process of growth by the agglomeration of matter, analogous to the growth of a rolling snowball, is called *accretion*.

36

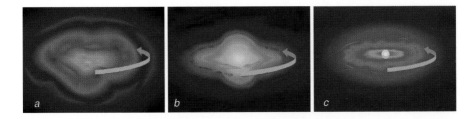

Artist's view of a collision between large planetesimals during the formation of the young solar system. Debris and dust clouds that would result from similar collisions may have recently been detected, in infrared observations, around a young star in the Pleiades cluster [37]. Credit: Lynette Cook.

Artist's view of a planetary system in formation. Credit: NASA.

The "pillars of creation." This portion of a nebula in the constellation of Aquila, also known as Messier 16 7000 LY from us, is a large cloud of dust and gas (mainly hydrogen) where many stars and planetary systems like ours may be assembling (Q. 157). Credit: NASA/ESA.

34 Is any trace of our "ancestral" supernova still in existence?

Planetary systems contain heavy elements that were synthesized during supernova explosions (Q. 16). These explosions disperse most of the star's material, leaving perhaps only a neutron star or a black hole behind. Has any such object been identified that may be the remains of the supernova that engendered our solar system?

The answer is no, and if any trace of it does still exist, it is probably impossible to identify. That hypothetical supernova detonated over 4.5 billion years ago (the age of our solar system), and this is actually a significant amount of time in the life of a galaxy. Since the explosion, our galaxy has made 20 to 30 revolutions. Meanwhile, the stars, even those orbiting at the same distance from the galactic center, have all continued to move with slightly different speeds and directions. So stars that were our neighbors 4.5 billion years ago, including the remnant of our "mother" supernova, are no longer near us, and there is no way to tell where it might be now.

35 How far out does our solar system extend?

When we call up a mental picture of our solar system, we imagine the Sun, the seven planets and their moons, the asteroids, Pluto, and now Eris. That list is far from complete.

The Solar System hosts many other bodies. For example, over 1000 Kuiper Belt objects are now known (out of a probable population of over 100 000). Located between 30 and 50 astronomical units (AU – see Q. 43) from the Sun, they are rocky and icy bodies, like the asteroids, or loose accumulations of rock and ice, like giant comets (Q. 69). Still further out is the Oort Cloud, the source of our long-period comets. The Oort Cloud may extend out as far as 100 000 AU (about 2 LY), and possibly harbors billions of rocky/icy objects of kilometer size (Q. 70). Despite their number, the total mass of all these small bodies is equivalent to only about three Earth masses.

The Oort Cloud is so far out that the Sun's gravity affects them weakly, and they can become influenced by the gravity of the nearest stars, which are only twice as distant. This far-off swarm of comets defines the borders of our solar system.

All the objects described so far are under the influence of the Sun's gravitational force, but there exists another domain beyond that: the extrasolar interstellar medium. The Sun produces a "wind" of electrically charged particles (Q. 47), and the pressure of this wind creates a sort of giant bubble whose limits are defined by a shock wave that marks the boundary between the region dominated by the solar wind, the heliosphere, and the interstellar medium beyond. The shape of this bubble is roughly spherical, with a radius of about 100 AU. The transition zone, which is identified by an

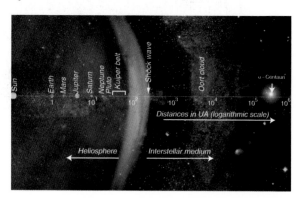

increase in the density of electrical particles, was mapped by the interplanetary probe Voyager 1 when it reached it in 2002, 25 years after being launched from Earth in 1977.

36 How old is the Sun?

Several different methods can be used to determine the age of the Sun. One method is based on the knowledge that the Solar System (Sun, planets, comets, etc.) formed as a coeval unit (Q. 33). The Sun is therefore at least as old as the planets, asteroids, comets, and other objects in the solar system. From the fossil record and the age of the oldest rocks, we know that the Earth is at least 4.55 billion years old (Q. 81), hence, the Sun must be at least that old.

A second method employs the theoretical model of star formation and nuclear combustion, which relates a star's age to its temperature, luminosity, and mass (Q. 11). This model has been verified for hundreds of stars of different types, including those very much like the Sun, which is a rather average star. It gives an estimated age of approximately 4.6 billion years for the Sun.

Finally, a recently developed technique called *nucleocosmochronology* allows us to determine the age of celestial bodies from the decay of their radioactive elements, a method similar to carbon-14 dating. For stars and other celestial bodies, the most useful elements are uranium or thorium, which have very long half-lives.[†] This is the most accurate method of all, and applying it to the Earth, meteorite, and the Sun itself gives an age of 4.57 ±200 million years.

37 Has the Sun always been as bright as it is now?

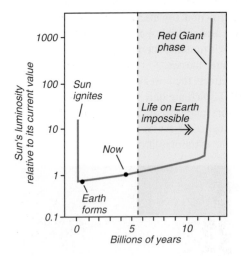

When nuclear combustion began in the core of the young Sun, the luminosity of our star shot up briefly to about 20 times its current level. Then it diminished and stabilized at about 70% of its present-day value. Since then, the luminosity has been gradually increasing and will continue to increase until the Sun enters the red giant phase (Q. 38) about 6.5 billion years from now. But things will go bad for us well before then (Q. 39).

[†] The *half-life* of a chemical element is the amount of time needed for half the atoms in a sample of that element to decay by losing protons from the nucleus. Some isotopes (Q. 88) have half-lives of years or thousands of years; other unstable isotopes may exist for fractions of a second to a few minutes.

38 What is our Sun's future?

The Sun is close to mid-life now. It will continue to shine normally for about another 6 billion years, although gradually increasing in volume and luminosity. At the age of 10.9 billion years, the Sun's central hydrogen reserve will be exhausted and our star will rapidly enter into a new phase that will trigger the burning of its central helium, inflating it into a red giant (Q. 15). Then, 1.7 billion years after that, with its helium exhausted and being unable to burn other elements, it will eject most of its envelope as a planetary nebula (Q. 157). The Sun will thus finish its life as a white dwarf (Q. 15).

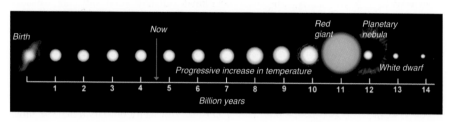

Evolution of the Sun (sizes are not to scale).

39 What will happen to the Earth when the Sun dies?

The scenario is catastrophic and inescapable. The Sun will shine normally for the next several billion years, but its temperature will steadily rise, making living conditions on Earth challenging well before our star's final convulsions. In approximately 1 billion years, the Sun will be 10% hotter than at present, and some think that this will destabilize the Earth's climate and biosphere, boiling off most of our water and transforming the surface of our planet into desert [11].

In 3.5 billion years the Sun's temperature will be 40% higher than now, and temperatures on Earth will be so high that any remaining oceans will have evaporated

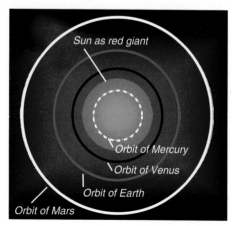

into space. The Earth will then share the fate of Venus: no water, no life. In 6.5 billion years, the Sun will inflate to become a red giant, triggering hellish conditions in the inner solar system. The Sun's outer layers will grow to encompass Mercury. Temperatures on Earth will reach 2500 °C, and melt the planet, transforming it into a ball of magma. As it continues to inflate, the Sun will suffer a significant loss of mass and, as a consequence, the orbits of the remaining planets will expand. Venus will be swallowed up, and though Mars may escape, it will not be habitable. The Earth is predicted to barely escape vaporization at the fringes of the red giant's surface. The giant outer planets (Jupiter, Saturn, Uranus, Neptune) are likely to survive this cataclysm, but will be modified significantly. After this final fiery gasp, our Sun will evolve into a slowly cooling white dwarf and the entire solar system will become very cold.

40 How hot is the Sun?

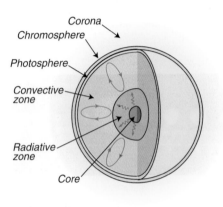

Temperatures in the Sun extend over a large range: between 15 million K at the core and 5800 K at the surface. Heat flows outward from the core towards the surface, but precisely how it flows is determined by how quickly temperatures change over radial distance, a change which is determined by local interactions between photons and the gaseous matter which constitutes the Sun. The restriction to radiation flow, called *opacity*, varies according to the composition, density, and temperature of the gas, and since the internal solar structure is complex, so are the heat-flow mechanisms and temperature distribution.

Heat generated in the core is in the form of gamma ray photons. At the high temperatures found there, the atoms are stripped of their electrons (i.e. the atoms are fully ionized) and cannot be raised to higher excitation levels when struck by photons, meaning that the photons are not absorbed. However, the numerous free electrons do bounce the photons around ("scatter" them), slowing down their advance. Temperatures along with density drop progressively with distance from the center of the core.

As we move away from the core, the temperature continues to drop, and eventually atoms become capable of holding onto some of their electrons. The hot plasma absorbs some radiation and the opacity increases for optical radiation. The temperature gradient is not very steep, however, and radiation is still the dominant heat transfer mechanism. This first shell, where radiation continues to dominate, is called the *radiative zone*. Its inner edge is at 7 million K and its outer edge is at 2 million K (see figure).

In the next region, opacity to radiation has become so high that another mechanism is needed to transfer the solar energy outward. Convection takes over at that point, with large convective cells now transporting the solar energy towards the surface – hot parcels of gas moving upward as cooler, denser parcels sink down to be reheated by the radiative zone. In this extremely thick layer, called the *convective zone*, the temperature falls from 2 million K at the inner edge to a surface temperature of 5800 K.

Above this zone, the density is so low that the layers again become transparent to radiation. This layer, the surface layer of the Sun called the *photosphere*, is at a temperature of 5800 K. From here, photons escape freely into outer space and it is this last layer that we see when we look at the Sun. The sudden change of opacity makes its edge look very sharp, giving the impression that the Sun has a distinct, crisp surface, but it is actually just gaseous, like the rest of the Sun.

The solar "atmosphere" has two important outer layers: the *chromosphere* and the *corona*. The chromosphere shines with such faint light that it cannot be seen against the bright photosphere; it can only be viewed during total solar eclipses, when the Moon masks the very bright solar disk. It has a pinkish color, hence its name (*chromos* means color in Greek). The structure of the chromosphere is dominated by *spicules*, highly dynamic jets of gas about 500 km in diameter and reaching 10 000 km in height. In the chromosphere, the temperature goes through a reversal, increasing with altitude from 4300 to 8300 K. This surprising temperature inversion is caused by the increased role of magnetic fields in transporting energy outward through the diffuse material.

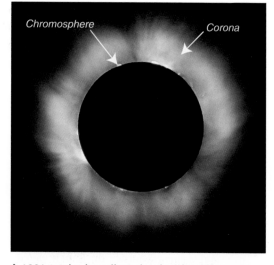

A 1991 total solar eclipse showing the chromosphere (in pink) and the corona (silver white). Credit: L. Viatour GFDL/CC.

The magnetic fields are also responsible for the very high temperature in the solar corona: 1 million K. Magnetohydrodynamic waves propagate in this medium and dissipate their energy there, like ocean waves crashing on the shore. Despite its high temperature,[†] the corona is not "hot" because its density is extremely low. The conditions are similar to those inside a hot kitchen oven: the walls are far too hot to touch, but you can put your hand inside and the hot air will not burn you. Even though it is at the same temperature as the oven walls, it is too tenuous to transmit much energy.

41 What causes sunspots?

Galileo, in 1609, was the first to study sunspots with a telescope (Q. 195)[‡] and his observations stunned his contemporaries. Since antiquity, the Sun had been considered a perfect sphere of fire – how could it have defects? But there they were, not only present but moving, day after day, from which Galileo correctly inferred that the Sun rotated on its axis about once a month.

Sunspots have been observed continuously since Galileo's time. We now know that they are short lived (lasting on average just a few weeks) and that their number and sizes change over time in an 11-year cycle. They are not dark but simply appear to be

[†] Remember that the temperature of a substance is a measure of the kinetic energy, or speed, of its atoms.

[‡] Sunspots had been seen before, by the Chinese in particular, as the larger ones are visible to the naked eye when the Sun is veiled.

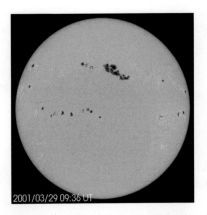

The Sun at its maximum of activity in 2001.
Credit: SOHO/ESA/NASA.

2001/03/29 09:36 UT

so, in contrast with the bright surrounding photosphere. If a sunspot could be extracted and examined outside the Sun, it would appear as bright as the full Moon. The lower luminosity of sunspots compared to the rest of the Sun is due to their lower surface temperature, 4500 K as opposed to 5800 K.

Sunspots arise from a complex interaction between solar magnetic fields and the Sun's *differential rotation*. Being gaseous, the Sun does not rotate as a solid body; instead, its rotation is faster at the equator (25 days) than at the poles (over 30 days). Solar magnetic field lines bundle in tubes – similar to the behavior of the jet streams in the Earth's atmosphere – and these tubes normally stay below the surface. Sometimes, however, the Sun's differential rotation distorts and stretches them, making them burst through the surface and producing what we see as a sunspot group. The strong magnetic field has the effect of slowing down the upward movement of gas in the sunspots, making these zones appear colder, hence less luminous.

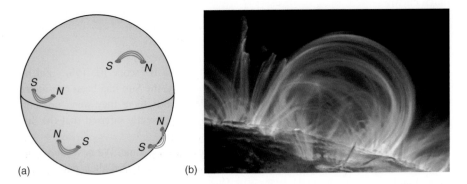

(a)   (b)

(a) Sunspots generally come in pairs or groups, linked by magnetic lines of force that emerge in one place and return to the surface at another. (b) Image of a post-flare loop prominence "mapping" the magnetic field lines (obtained by the satellite TRACE in 1999). These loops are immense, 30 times the size of the Earth, measured side to side. Credit: NASA.

Surface motions can sometimes drastically distort the Sun's magnetic field lines of force. The lines may be sheared off, then reconnect, causing a kind of short circuit that produces a solar flare, or eruption.

42 Do sunspots influence the weather on Earth?

Sunspots have been systematically observed since the invention of the telescope 400 years ago. Their study has revealed that they evolve over a cycle of 11 years, although there are other long-term variations, for example the Maunder Minimum, which lasted from 1645 to 1715, when essentially no sunspots were visible. This period also corresponded to the "Little Ice Age" during which the climate in Europe was exceptionally cold, suggesting that there may be a link between Earth's climate and solar activity. Nonetheless, no clear mechanism for a possible relation between climate and solar activity has been put forward.

The true length of the Sun's magnetic activity cycle is 22 years, not 11, because there is a reversal of the solar magnetic poles every 22 years at the time of maximum activity. The last maximum of solar activity took place in 2000, and the next one will occur in 2011. The last minimum was in 2007.

The solar magnetic field produces not only sunspots, but also jets of ionized gas that are associated with *flares*, and which can affect the atmosphere of Earth. When this happens, *aurorae* (northern/southern lights – see Q. 118) ignite the sky of high latitude regions on Earth, disturbing radio communications and sometimes affecting high-voltage electrical transmission lines. In 1989, a blackout induced by the Sun left six million Quebeckers without electricity.[†]

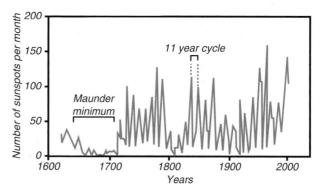

43 How was the distance to the Sun measured?

In 1769 Captain James Cook, on orders from the London Royal Society, traveled to a pretty promontory in Tahiti, French Polynesia, now called "Pointe Vénus," to observe the "transit of Venus," the phenomenon of Venus passing across the face of the solar disk. His mission: to provide a measurement that would help determine the size of our solar system.

[†] Protective measures have since been adopted to reduce the vulnerability of long high-voltage electrical power lines to the effects of the severe subsurface currents generated by strong solar flares.

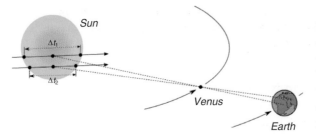

By measuring the duration of the Venus transit from two distant points on Earth, the absolute distance between the Earth and the Sun can be derived.

This had become the key problem of eighteenth-century astronomy. By using Kepler's laws, the relative positions of the planets had already been determined, but their absolute distances were inaccurate. The few triangulation measurements that were available were very coarse; it was like having a map of a country with all the towns marked, but no scale for judging distances.

The transit of Venus was a chance to establish the scale of the solar system by establishing the Earth–Sun distance using the direct parallax method (Q. 6). It was a rare opportunity, since Venus crosses the solar disk only twice per century. Although England, Austria, and France were at war at the time, their scientific institutions decided to coordinate efforts to observe the transit from several distant points on Earth (the longer the baseline, the more accurate the measurement). Many astronomers traveled to far-flung locations in Europe, Africa, North America, and Polynesia to observe the event.

Today, the most precise way to measure the distance to Venus (or any planet) is by radar; the distance to the Sun is then derived by trigonometry. This distance, which proves to be about 150 million km, is now known to a precision of within ± 30 m. The Earth–Sun distance is the cosmic yardstick of astronomers. Called the astronomical unit (AU), (Q. 7) it is the basis of all measurements of distance to stars and to the galaxies outside our Milky Way (Q. 6).

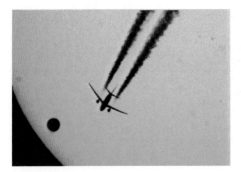

The transit of Venus in June 2004, viewed from the Netherlands. By chance, a plane became part of the spectacle. Credit: Koschny/ESA.

44 Is the distance between Earth and the Sun changing?

The distance from the Earth to the Sun changes in a periodic way by about 3% throughout the year, since the Earth's orbit is slightly elliptical. However, two systematic effects have the Earth slowly drifting away from the Sun. The first is a tidal effect with the Sun, like that acting on the tandem Earth–Moon (Q. 105). This effect is minute; the increase in the distance to the Sun is only 0.1 mm per century. The second is due to the Sun losing mass! The solar luminosity is generated by the thermonuclear fusion of hydrogen into helium that results in a mass annihilation after Einstein's relation $E = mc^2$ (Q. 1). As the gravitational attraction decreases with time, the Earth's orbit expands. This effect is also very small: about 1 cm per year.

45 How can we know the mass of the Sun?

Newton's gravitational law gives us the answer. The Earth remains in its orbit because the centrifugal force pulling it outward is balanced by the attraction of the Sun (Q. 106). Equating these two forces gives us the mass. Assuming the Earth's orbit to be circular, the centrifugal force is given by $F_c = v^2/R$, where v is the orbital velocity of Earth around the Sun (derived from the time it takes the Earth to orbit the Sun, i.e. one year), and R is the distance from the Earth to the Sun. As for the attractive force between Earth and the Sun, it is given by Newton's law $F = GmM/R^2$, where G is the constant of gravity, m is the mass of the Earth, and M is the mass of the Sun (Q. 80). Equating the Sun's attractive force and the centrifugal force, and knowing G, m, and R (Qs. 80 and 43), provides M, the mass of the Sun, which is found to be $2 \cdot 10^{30}$ kg, or 330 000 times that of the Earth.

The mean density of the Sun is a third that of Earth (or 1.4 times that of water). Most of the mass of the solar system is concentrated in the Sun; the collective mass of the planets, asteroids, comets, and dust in the solar system is barely 0.2% that of the Sun.

46 What is solar radiation pressure?

Solar radiation – or light – is composed of photons. A surface that is being illuminated by the Sun is simply being bombarded by photons, and those photons exert a pressure. This is a fundamental effect; all electromagnetic radiations cause pressure, radiation pressure.

Photons are not like bullets: they have no mass. How can they exert pressure if they are massless? While they have no mass when at rest (which is meaningless, since light is never at rest), photons do have energy and momentum,[†] and the latter translates into a pressure when light strikes an object. Radiation pressure is very weak, but with large surfaces, its effect is no longer negligible.

Solar radiation pressure produces the blue tails in comets (Q. 72), and it may affect the orbit of planets. It introduces perturbations on large-sized satellites. It could also be employed for the propulsion of spacecraft equipped with what are called "solar sails" (Q. 77).

47 What is the solar wind?

Solar wind is the name given to the supersonic stream of solar particles (protons, electrons, and other ions) ejected by the solar corona (Q. 40) which completely floods

---

[†] Momentum is the product of the mass of a body and its velocity: $p = mv$. In relativistic conditions, mass and energy are equivalent according to the relationship $E = mc^2$: the photon may be considered to have a virtual mass of $m = E/c^2$, and its momentum is then $p = mc = E/c$. That momentum is imparted on the object with which photons come in contact. If U is the power received from the Sun per square meter, the radiation pressure is U/c. At the distance of the Earth, the solar radiation pressure is five micropascals.

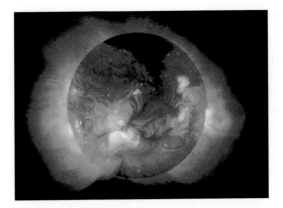

The Sun imaged in x-rays by the spacecraft Yokoh in 1992. The darker regions are the coronal holes, the main regions where the solar wind is produced. Credit: Yohkoh, Japan.

the solar system. These particles escape from the gravitational field of the Sun because of the very high temperature in the solar corona (more than a million K), and are accelerated by a nozzle-like effect to very high speeds, 400 km/s. They can reach the Earth in about four days.

The solar wind originates mainly from zones on the Sun called *coronal holes*, where the magnetic field lines open to outer space. The holes are located mainly at the poles. Although the solar wind carries away about 1 million ton/s, this represents a mass loss of less than 0.1% since the birth of the Sun.

The solar wind interacts with the atmospheres (or surfaces) and magnetic fields of planets; in the case of Earth, it may trigger magnetic storms or aurorae (Q. 118). It is the cause of one of the two types of comet tail (Q. 72).

48 How long does light from the Sun take to reach us?

Sunlight takes about 8.3 min to reach us. This is the amount of time needed by light to travel the distance to Earth 150 million km at the speed of 300 000 km/s. Since the Earth's orbit is elliptical, the time varies slightly. The time lag is 8.1 min during the northern winter (when the Earth is closer to the Sun) and 8.45 min during the summer (in the northern hemisphere).

49 What is the difference between a star and a planet?

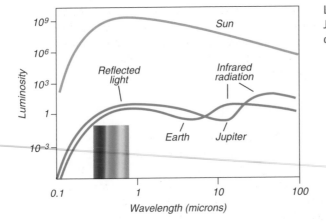

Luminosities of the Sun, Earth, and Jupiter when observed from a distant star.

At first glance, the difference seems obvious: a star is a celestial body that does not belong to the Solar System (except for the Sun), and that is self-luminous. A planet is a body in orbit around the Sun and that reflects its light. This can be generalized: "exoplanets" are non-luminous bodies in orbit around other suns.

If we delve a little deeper, though, things become more complicated because planets, too, are self-luminous – all objects with a temperature radiate. Planets radiate primarily in the infrared (because their temperatures are much lower than stars') and hardly at all in the visible. The Earth radiates into space as a body at 0°C (273 K) – "hot" compared to the temperature of space (slightly below 3 K) – and can easily be detected from space with an infrared camera. Venus, Jupiter, and the other planets also emit in the infrared and at radio wavelengths. If Jupiter had accreted more hydrogen during its formation, it could have maintained nuclear fusion for some time and would by now have evolved into a brown dwarf (Q. 50).

The distinction between star and planet can thus be fuzzy. Nevertheless, let us say that a star is a body that is self-luminous from the generation of energy by thermonuclear reactions, whereas a planet is a relatively large body in orbit around a star, and that it shines by its own thermal emission and light reflection from its parent star. In addition, planets host a range of elements in gaseous, liquid, and solid form (rocks, ice, minerals, water, etc.) while stars are totally gaseous (most of their life, at least).

50 What is a brown dwarf?

Although there are no brown dwarfs in the solar system, it is relevant to mention them in the context of the previous question; brown dwarfs are intermediate between stars and massive Jovian planets. Their core temperatures and pressures are not high enough to maintain hydrogen fusion because their mass is too low (more than 10 times that of Jupiter but less than 7% the mass of the Sun). They are like "aborted" stars, in a sense, but self-luminous, generating heat from their slow gravitational contraction. The more massive brown dwarfs enjoy a brief period of thermonuclear activity, burning deuterium (or lithium), but they quickly exhaust their limited supply of fuel, never reach a stable state, and quietly go "extinct."

As the luminosity of brown dwarfs is very low, they are difficult to find. The first one was discovered in the early 1990s, and by now several hundreds have been identified. There are a colossal number of them in the Milky Way. Brown dwarfs may exist in isolation (the so-called *free floaters*), or orbit around normal stars. The coolest ones have temperatures of only a few hundred K. Complex atmospheric phenomena such as those found on planets occur in their outer layers.

51 Why are some planets rocky and others gaseous?

The "rocky" planets in our solar system, Mercury, Venus, the Earth, and Mars, are composed mainly of metals and rocky minerals and are endowed with little or no atmosphere. The Moon – which is almost a planet-sized body – has no atmosphere at all. Conversely, the gaseous planets, Jupiter, Saturn, Uranus, and Neptune, are almost entirely composed of atmosphere (most of their mass resides there) while their rocky cores are proportionately small. Since all the planets formed from the same primordial

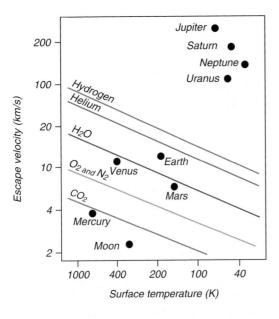

solar nebula containing mostly hydrogen and helium, one would expect that every planet would have a solid nucleus built from planetoids surrounded by an atmosphere of hydrogen, helium, and other gases. What happened?

The culprit was the Sun. The rocky planets were unable to accrete and retain their hydrogen and helium because of their proximity to the Sun. Remember: the lighter a gas is and the higher the temperature, the more rapid is the movement of the gas atoms. When they are moving fast enough, they escape from the gravitational field of their planet and off they go, shooting out into space. The critical speed to take flight, called the *escape velocity*, depends on the surface gravity of the planet.[†]

The figure shows the different planets, selected atmospheric gases, and the conditions for a gas to escape or be retained as a function of escape velocity and temperature. Above the lines, a gas escapes; below, it remains bound by gravity. For the rocky planets (Mercury, Venus, Earth, and Mars), temperatures are too high and the planetary mass too small to retain hydrogen and helium, whereas Jupiter, Saturn, Uranus, and Neptune are cold and massive enough to retain these gases.

It is interesting to note that although the major constituents of the massive planets (hydrogen, helium, ammonia, and methane) are gaseous under normal Earth conditions, they are liquid under the huge pressures that prevail in the interiors of giant planets. The outer planets of our solar system might best be described as "liquid planets."

52 What are the interiors of planets and satellites like?

The "rocky" planets (Mercury, Venus, Earth/Moon, and Mars), satellites, and the solid cores of the gaseous planets are structured in layers: at the center there is a metallic core (mainly iron), then usually some intermediary layers, and finally a rocky crust. The building blocks for the planetary bodies were planetesimals (Q. 33), which are relatively homogeneous mixes of metals and silicates, and one might reasonably expect them to produce homogeneous planetary interiors. That is not the case, however. The interiors are stratified. Here is what happened.

[†] The escape velocity is given by the formula $v_l = \sqrt{2GM/R}$, where G is the gravitational constant, M the mass of the planet or satellite, and R, its radius, or also $v_l = \sqrt{2gR}$, where g is the acceleration at the surface of the body.

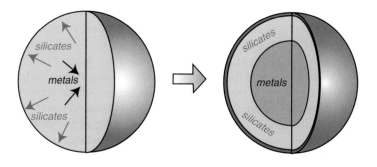

As the planetesimals coalesced, heat was generated through several processes:

• collisions during accretion;
• compression of the large planetoids by gravity;
• radioactive decay (the case of Earth);
• "kneading" of the crust by tidal effects (the case of Europa, one of Jupiter's moons).

The nascent planets partially melted and the viscous materials became segregated according to their density, just as oil and vinegar separate into distinct layers in a salad dressing. The heavy elements (metals) sank to the center, and the lighter ones (silicates) migrated to the surface, a phenomenon called "planetary differentiation." As the bodies cooled, the lighter materials solidified to form the rocky crusts of planets and satellites.

53 Where do the names of the planets come from?

The Greeks and Romans bequeathed to us the names for all planets that are visible to the naked eye (Q. 189).

Venetia Burney, who proposed the name for Pluto.

Mercury was Jupiter's messenger and the protector of tradesmen. The planet was probably named after him because of its speedy movement across the sky.

Venus, the goddess of beauty and love, gave her name to the most spectacular planet in the evening or morning sky.

Mars was the god of war. The red color of the planet probably reminded the ancients of the color of blood. In mythology, Mars had two sons by Venus: Phobos and Deimos (Fear and Panic). Those names were given to the two satellites of Mars when they were discovered in 1877.

Jupiter, the god of the heavens, lightning, and thunder, reigned over the other gods. The planet was probably named after him because of its brightness and slow, majestic movement across the sky.

Saturn was the original king of all the Roman gods, but he was dethroned by his son, Jupiter. The planet that bears his name is less bright than Jupiter and has a longer period of revolution around the celestial sphere.

Uranus, Neptune, and Pluto were discovered only after telescopes came into use. In naming them, the tradition of using personages from Greek and Roman mythology was retained: Uranus was the primordial Greek god of the heavens and Neptune the

Roman god of the sea. Not that the tradition always went uncontested: Herschel, who discovered Uranus, wished to call it *Georgian Planet* in honor of his patron, King George III, and Arago, the director of the Paris Observatory, lobbied to call Neptune *Le Verrier*, to honor his colleague who had predicted its existence and position. As for Pluto, it was an 11-year-old English girl who had the honor of naming this one. Fascinated by mythology, she suggested the name of the god of the dead because of the planet's ability to make itself invisible. Her grandfather, a librarian at Oxford University, sent her proposal to the American astronomer, Percival Lowell, who had predicted the planet's existence (Q. 244), and the name pleased him – not least because Pluto began with his own initials ...

54 What is Bode's law?

Planets are often represented neatly lined up and equidistant from each other. In fact, the distances between their orbits increase as one moves out from the Sun.

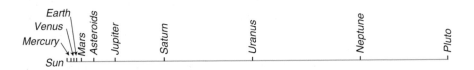

In 1766 the German astronomer, Johann Daniel Titius, noted that the semi-major axis of the planetary orbits obeyed a simple numerical law. Six years later, Johann Elert Bode, director of the Berlin Observatory, published the formula without mentioning Titius, and the law became known as Bode's law. Now, however, in due recognition of its discovery by Titius, it is more properly called the *Titius–Bode law*. It goes as follows (see table):

We start with the numerical series 0, 3, 6, 12, 24, 48, 96, ... ; except for the first two values, each number is twice the value of the preceding one. Next, we add 4 to each number, giving 4, 7, 10, 16, etc. Finally, we divide by 10. The results give the semi-major axis of the classical planets (Mercury to Saturn) expressed in astronomical units (the radius of Earth's orbit), with a missing planet between Mars and Jupiter.

Few astronomers paid much attention to Bode's Law until, in 1781, interest in it was piqued by the discovery of Uranus and the realization that this planet, too, had an orbit that satisfied the law. Bode then proposed that a search be made for a potential planet between Mars and Jupiter, in the zone predicted by his law. This led to the discovery of Ceres in 1801. Ceres, whose orbit is in the main asteroid belt, is now one of the larger representatives of the minor planets.

Planet	Sequence	Bode's law	Actual value
Mercury	0	0.4	0.39
Venus	3	0.7	0.72
Earth	6	1.0	1.00
Mars	12	1.6	1.52
Ceres	24	2.8	2.77
Jupiter	48	5.2	5.2
Saturn	96	10.0	9.4
Uranus	192	19.6	19.2
Neptune	384	38.8	30.1
Pluto	768	77.2	39.4

When Neptune was discovered, however, the law failed. Curiously, it is Pluto that is closest to the predicted value for a body beyond Uranus. Numerous other bodies that do not follow the law, including Eris and the Kuiper Belt objects, have since been found.

Bode's law is now considered something of a historical curiosity. It is probable that the cases where the law applies correspond to the well-known resonances of planetary orbits, but no theory or simulation has yet been able to prove it. The absence of a true planet between Mars and Jupiter could be due to orbital resonances and tidal effects from Jupiter that would have prevented the coalescence of asteroids in this zone.

55 What is Planet X?

For nearly a century and a half after the discovery of Uranus by Herschel in 1781, its unorthodox motion around the Sun intrigued astronomers. Many thought that the puzzling anomalies were due to the gravitational influences of nearby bodies. Eventually, a detailed analysis of its orbit led Le Verrier to discover Neptune in 1846. That did not explain all the vagaries of Uranus's orbit, however, and the American astronomer Percival Lowell later carried out a systematic search for a planet beyond Neptune, calling it "Planet X" (X for unknown, as in algebra). This led to the discovery of Pluto in 1930 by Clyde Tombaugh (Q. 244). But again, Pluto had too small a mass to explain the remaining anomaly in Uranus's orbit.

Today, we no longer need to posit that elusive Planet X to explain away the mystery. Voyager 2's 1989 flyby of Neptune provided us with new data according to which the mass of that planet proves to be about 0.5% less than previously thought; all the calculations have been redone and there are no remaining discrepancies. Furthermore, none of the spacecraft that have traveled to the outer solar system (Pioneer 10, Pioneer 11, Voyager 1, and Voyager 2) have exhibited any peculiarities in their trajectories.

If the need for the "Planet X" of old is gone, however, the same term is sometimes used today when astronomers speculate about the existence of a cold dark planet in the outermost reaches of the solar system. Alas for mystery, if any such body exists, it should have been detected by the spacecraft IRAS that completed a full infrared map of the sky in 1983. Nothing was found then, nor during several other recent infrared surveys. Interestingly, in 1992, the first Kuiper Belt object was discovered (Q. 69), and thousands more have been found since then. The Kuiper Belt has a distinct outer boundary at approximately 50 AU. What might explain that? Might there not be a planet the size of the Earth out there, one that swept the outer part of the Kuiper Belt the way the moons of Saturn create distinct gaps in that planet's ring system?

56 Why is Pluto no longer a planet?

In 2006, when new criteria were adopted by the International Astronomical Union to define a planet, Pluto was demoted from the ranks. With a diameter of only 2306 km, it is smaller than the Moon and many satellites of the outer planets (Europa, Callisto, Ganymede, Io, Titan, and Triton).

The term planet is now restricted to a body in orbit around the Sun whose mass is sufficient to have eliminated by accretion all other objects moving in orbits close to its

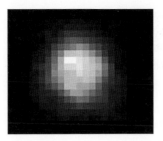

Pluto imaged with the Hubble
Space Telescope in 1994.
Credit: NASA.

own. Because of its low mass and highly inclined elliptical orbit, Pluto failed that test.
It now belongs to a new category, the "dwarf planets." Members of this group include
the asteroid Ceres, and Eris, the larger body found beyond Pluto in 2003. There is also a
third category of objects, the "small bodies," that are in orbit around the Sun but are not
massive enough for gravity to pull them into a spherical shape (Q. 3), a characteristic
that applies to most asteroids, comets, and Kuiper Belt objects (Q. 69).

Pluto is difficult to observe. Only the Hubble Space Telescope and the largest ground-
based telescopes equipped with adaptive optics system have been able to make images
of its disk. It is the only (ex-)planet not to have been visited by a spacecraft. However,
the probe *New Horizons*, launched in 2006, will reach it in 2015, before continuing
on out to explore the Kuiper Belt. Despite being very small, Pluto has a suite of three
moons: Charon (diameter 1200 km), discovered in 1978, and two other small satellites,
Nix and Hydra, recently discovered by the Hubble Space Telescope.

57 Why do some planets have many satellites and others, none?

The rocky planets (Mercury, Venus, Earth, and Mars) and the dwarf planets (Pluto
and Eris) have few satellites (Mercury and Venus have none), while the four giant
planets (Jupiter, Saturn, Uranus, and Neptune) have hundreds of them. There is an
unquestionable trend in which the number of satellites increases with the mass of

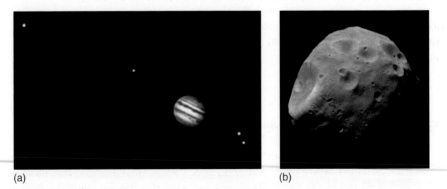

(a)   (b)

(a) Jupiter's four Galilean satellites. (b) Phobos, Mars's satellite, which resembles
a captured asteroid (image taken by the Mars Express probe in 2004). Credit: NOAO
and ESA.

the planet. This is not surprising since a larger planet in formation will more easily be able to attract nearby bodies and later capture flyby objects. It is likely that a large fraction of the 63 and more known satellites of Jupiter were stolen from the asteroid belt and, similarly, that many of Neptune's 13 satellites were abducted from the Kuiper Belt.

58 How can Mercury survive so close to the Sun?

Mercury is about twice as close to the Sun as Earth, and the intensity of the sunlight striking it is seven times as strong. During daytime, at the equator, a human being would be burnt to a cinder, as the temperature reaches 420 °C.

The nights are cold (−170 °C) for two reasons. First, there is no atmosphere to redistribute the heat between the side facing the Sun and the opposite, nighttime side. Second, a "day" on Mercury is very long: the equivalent of two of our months. This allows the nighttime side to cool down.

Mercury's South Pole, where the floors of some craters could be covered with ice (photograph by the probe Mariner 10 in 1974). Credit: NASA.

Although Mercury's surface is subject to serious thermal blasting, the temperature during the day is not high enough to melt the surface rocks. This might have been the case in the distant past, however. Under the heat of the young Sun, the surface temperature on Mercury may have gone as high as 3000 °C. A good fraction of the crust would then have evaporated and been swept away by the solar wind. This would explain why Mercury's rocky crust, when compared to the size of its iron core, is thinner than for the other planets.

Despite the extremely high surface temperatures, radar observations indicate that there may be water hiding at the poles in the form of ice. This surprising possibility stems from the fact that sunlight never reaches the bottoms of some of Mercury's polar craters.

59 Why does Venus have phases like the Moon?

Venus and Mercury show phases like the Moon because they are "inferior" planets; that is, their orbits are located between Earth and the Sun. When viewed from Earth, they are always close to the Sun and are visible only within a few hours just before sunrise or after sunset. When we see them, their disks always appear partly lit, i.e. show phases, as in cases (a) and (c) in the figure. We do not see them when they are between us and the Sun (case (b)), presenting their dark sides to us, nor can they be seen when they are on the far side of the Sun with respect to Earth (case (d)), as this corresponds to daytime for us and their disks are drowned in the bright sunlight.

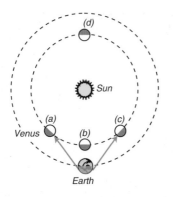

Mars and the other "superior" planets with orbits outside ours almost always appear fully lit, and do not have phases.

60 What is the Great Red Spot on Jupiter?

The Great Red Spot of Jupiter photographed by space probe Voyager I in 1979. Credit: NASA.

The Great Red Spot is a giant hurricane in the atmosphere of Jupiter. It has been raging for at least 350 years, with winds up to 500 km/h. On Earth, hurricanes last for several weeks on average, weakening and fading away when they reach continental land where the evaporation of ocean waters is no longer supplying them with energy. On Jupiter, there are no land masses and, once a huge vortex has formed, it goes on and on. Jupiter's atmosphere is made essentially of hydrogen and helium, with small amounts of methane and ammonia. Since none of these gases is colored, the origin of the striking colors of the planet's atmospheric bands and Great Red Spot remains a mystery.

61 What are Saturn's rings made of?

When viewed from Earth, Saturn's rings look quite solid, but they are actually made up of a myriad of icy and rocky pebbles ranging in size from a few centimeters to a few meters. The gaps between the principal rings are either caused by satellites that have swept these spaces out (i.e. attracted the material and incorporated it into themselves), or by dynamic resonances that keep the gaps clear of material.

Each of the principal rings is made up of a multitude of mini-rings. The rings are very narrow (of the order of 1 km) and thin (some only 10 m thick), while their diameters

The interplanetary probe Cassini captured this dramatic image of Saturn in 2007. The exposure was set to emphasize the rings, hence the globe of Saturn appears overexposed. Credit: ESA/NASA.

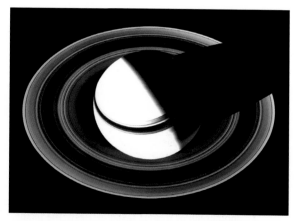

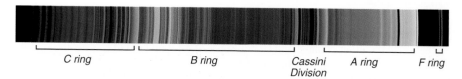

| C ring | B ring | Cassini Division | A ring | F ring |

Mosaic of photographs from the probe Cassini showing the complexity of the rings. Dust and chemical compounds mixed with ice in various combinations are probably responsible for their subtle range of colors. Credit: ESA/NASA.

span over one million km. If assembled into a single body, they would barely be the equivalent of a mini-moon 100 km in diameter.

Observations from the spacecraft Voyager suggested that the rings were the broken remnants of a large ancient moon of Saturn, possibly one that was hit by a comet. Collisions between the rocky and icy fragments of the rings dissipate their kinetic energy, guaranteeing that, in the long term, the rings will disappear. Their lifetime is estimated at about 100 million years. New observations from the *Cassini* probe, however, indicate that the rings are being continuously reborn from recycled fragments and cosmic dust.

Saturn is not unique in having rings. Jupiter, Uranus, and Neptune have them, too, although those systems are more diffuse.

62 Do all the planets orbit in the same direction?

Imagine yourself floating in outer space far above the Earth's North Pole. Looking down, you would see all of the planets revolving around the Sun in almost exactly the same plane and in the same direction – counterclockwise – which is also the direction of the Sun's rotation. You would also see almost all of the planets rotating in the same direction and with their axis close to perpendicular (within 30°) to the plane of their orbits. This is consistent with the scenario wherein all the planets formed from the same solar nebula that gave birth to the Sun (Q. 33); its "fossil" movement has remained imprinted on all bodies.

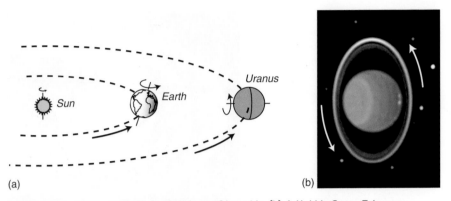

(a)   (b)

(a) Uranus's axis of rotation is in the plane of its orbit. (b) A Hubble Space Telescope photograph of Uranus taken in 1997. The image, in false colors, shows the sunlit pole on the left. The planet, its ring, and its satellites revolve in the direction shown. Credit: Karkoschka/NASA.

There are two exceptions: Venus and Uranus. Venus rotates in the direction opposite to that of all the other planets. And Uranus's axis of rotation (and of its satellites) points strangely into the plane of its orbit. What happened? The most plausible scenario is that these two planets experienced major impacts at the time of the formation of the solar system, when collisions were frequent, knocking them out of alignment.

63 What are the Lagrangian points?

The trajectory of two bodies orbiting each other in space can be fully determined using Newton's gravitational theory: they will follow elliptical orbits around their common center of mass. If a third body is introduced, however, the situation is no longer so simple. The trajectories are no longer closed curves, and they cannot be described by simple equations such as Kepler's laws. This is the famous "three-body problem" which has stumped the best mathematicians of the last several centuries.

A very specific solution does exist, however, when the mass of the third body is negligible with respect to that of the other two, and when the object of intermediate mass (e.g. the Earth) is in a circular orbit around the most massive one (e.g. the Sun). In this case there are five points in the orbital plane where a body of low mass would move in a circular orbit synchronized with that of the body of intermediate mass, because, there, the combined gravitational attraction of the two large bodies is exactly balanced by centrifugal force. These privileged locations are called "Lagrangian points" and are referred to as L_1, L_2, ..., L_5. Each pair of large mass bodies in the solar system has such a system of points: the Sun has one with each of the planets, as does the Earth–Moon system.

Two Lagrangian points are stable: L_4 and L_5. A body arriving at one of them can be captured and will remain there. For example, approximately 1000 asteroids, called the *Trojans*, are "parked" at these points in the Sun–Jupiter system. The Sun–Neptune system has its own "herd," too. The Sun–Earth couple does not have one, but dust accumulations have been observed at those points.

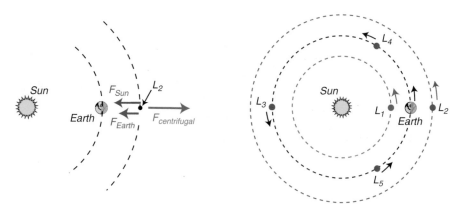

At a given Lagrangian point of the Sun–Earth system (here L_2 on the drawing, left), the centrifugal force is equal to the gravitational attraction of the Sun and the Earth. Right are shown the five Lagrangian points of the Sun–Earth system that revolve around the Sun in synchronism with the Earth. L_1 and L_2 are located well beyond the orbit of the Moon.

The three other points, L_1, L_2, and L_3 are metastable: an object that arrives there will escape if perturbed, and in space, perturbations are continuous, in particular from solar radiation pressure (Q. 46) and the gravitational attraction of nearby objects. Nevertheless, objects placed at a metastable point drift out only slowly, and an artificial satellite can be kept there through careful orbit maneuvering. L_1 and L_2 are particularly advantageous: L_1 for continuous observation of the Sun or of the sunlit face of Earth, and L_2, conversely, to avoid seeing them. The solar observatory SOHO has been stationed at L_1 since 1995, and many astronomical observatories are already positioned, or are scheduled to be positioned, at L_2 (Q. 223 and 224).

64 Why did the comet Shoemaker–Levy 9 break up as it approached Jupiter?

Small bodies such as asteroids and comets are composed of particles that have coalesced under their mutual gravitational attraction. When such a body approaches a planet, the pull of gravity engenders tidal forces (Q. 105) that may overcome the cohesive forces holding the particles together. The small body shatters.

This is precisely what happened to comet Shoemaker–Levy 9 in 1994; it fragmented into

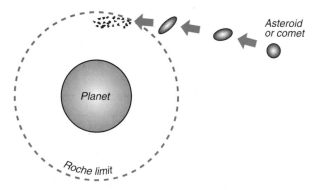

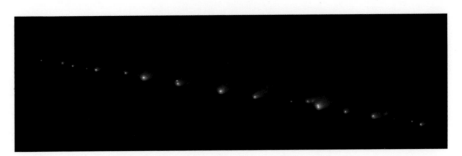

Multiple fragments of comet Shoemaker–Levy 9 near Jupiter. Credit: Weaver and Smith, STScI/NASA.

24 pieces, torn apart by Jupiter's gravitational field. This is the same mechanism that helps perpetuate the rings of Saturn (Q. 61).

The distance at which such a breakup occurs is called the *Roche limit*, after the French mathematician Edouard Roche (1820–83) who established the concept in the mid-nineteenth century. The Roche limit depends on the density of the planet and of the small body. When these densities are identical, the limit is reached at 2.4 times the radius of the larger body.

The concept of the Roche limit can only be applied to objects whose cohesion is due to the gravitational attraction between their constituent particles. This is not the case for bodies that melted and solidified, such as a block of rock or of ice, because the cohesion there is due to chemical bonds between atoms. Naturally, the Roche limit does not apply to space probes either, since they are also made of fused material and are structurally consolidated.

65 Can planetary alignments cause catastrophic events on Earth?

In March 1982, Mars, Jupiter, and Saturn appeared in the night sky within a few degrees of each other in an unusual conjunction. Uranus and Neptune also happened to be on the same side of the Sun as the Earth then, although were somewhat less neatly aligned. The spectacle in the night sky was magnificent – but that is not why it made news headlines. That happened because, several years previously, two respected astronomers had predicted that the alignment would trigger natural catastrophes on Earth due to the combined attractive forces of the planets and the Sun. The great San Andreas fault in California would suddenly slip and Los Angeles would be destroyed, for instance [23]. The fact that the astronomers had retracted their prediction fully two years before the conjunction occurred did nothing to prevent the prediction from being used to stir up public emotion and sell newspapers, of course. Naturally, nothing very dire happened when the conjunction finally took place. A similar alignment occurred in 2000, also to no particular effect.

It is quite true that tidal forces can sometimes shatter a body that approaches another too closely. This happened to comet Shoemaker-Levy 9 when it flew near Jupiter in 1994 (Q. 64). However, tidal effects produced by the other planets are totally negligible on Earth.

Alignment of the Sun, Earth, Mars, Jupiter, and Saturn in 1982.

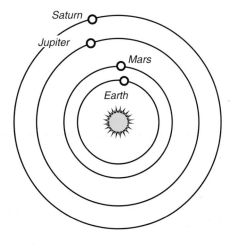

The results tabulated here are convincing. The second column shows the values for the gravitational forces exerted on the Earth by other planets (normalized to that of the Sun), as derived from Newton's second law, which states that the force of gravity decreases with the square of the distance. The Sun obviously exercises the greatest force, but it is not the gravitational attraction itself that is really relevant here. What matters is the *difference* in its value between the two opposite sides of Earth. It is this difference in value between one side of the Earth and the other that makes the ocean levels rise in tides and would tend to make the solid crust of Earth shift. This difference (the tidal force), given in the third column (normalized to the tidal force of the Sun), decreases more rapidly with distance than does the gravitational force, since it is a function of

	Attraction	Tidal force
Sun	1	1
Moon	$5.1 \cdot 10^{-3}$	1.9
Venus	$3.2 \cdot 10^{-5}$	$1.2 \cdot 10^{-4}$
Mars	$1.2 \cdot 10^{-6}$	$2.2 \cdot 10^{-6}$
Jupiter	$5.2 \cdot 10^{-5}$	$1.3 \cdot 10^{-5}$
Saturn	$3.8 \cdot 10^{-6}$	$4.5 \cdot 10^{-7}$

the cube of the distance, not of the square (Q. 106). Although the gravitational attraction of the Moon is 373 times weaker than that of the Sun, its tidal force is 1.9 times greater because it is closer. Looking at the values for the other planets, we see that their influences are minute; the tidal force from Venus is 15 000 times weaker than that of the Moon. The tides from the other planets are insignificant.

When next you have the opportunity to see a spectacular planetary conjunction, just enjoy. There is nothing to fear!

66 Did asteroids cause the mass extinctions on Earth?

When an animal or plant dies, its remains are often eaten by other animals or destroyed by bad weather or erosion – or, again, it may by chance fall into a pond or lake and sink into the mud, which protects it from decaying. With time, it may fossilize and be covered by additional layers of sedimentary rock. By studying these rock strata, the

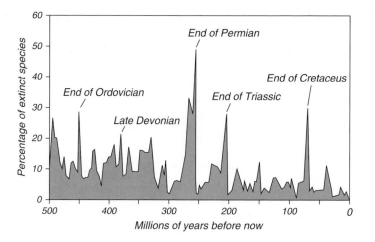

species that lived at the different epochs of the Earth's history can be reconstructed and dated.

Sometimes, fossils of a species abundant in a given layer no longer appear in the layer just above it or, conversely, new species may appear there. Ruptures in continuity of this sort often indicate extinctions – in some few cases, massive ones involving 30 to 60% of all species on Earth. Discontinuities can take place over surprisingly short periods, a few million years, the blink of an eye on the geological scale.

The extinction that took place at the end of the Permian period 250 million years ago was particularly catastrophic: 95% of all marine species disappeared, along with 70% of terrestrial vertebrates. The end of the Cretaceous period 65 million year ago witnessed the extinction of the dinosaurs. How can such gigantic upheavals be explained?

A number of different causes have been put forward: unusually intense continental drift, massive volcanic eruptions, significant changes in the average global temperature, etc. A sizable asteroid smashing into Earth could also produce a cataclysm of the required magnitude. Much evidence actually supports such a theory to explain the extinctions at the end of the Cretaceous period (Q. 67). The impact of an asteroid in Australia has also been proposed to explain the great extinction of the late Permian [4].

Although the impact theory for major extinctions still awaits complete confirmation, the concept is a very plausible one. Based on the frequency of cratering events on the Moon, it has been calculated that Earth has been struck by a body in the 10 km size range about once every 100 million years. An impact by an object of that size produces the equivalent of 5 billion thermonuclear bombs, projects huge quantities of dust into the upper atmosphere, and creates dark cosmic winters lasting decades. Who could doubt the calamitous effects that such events would have on the world's fauna and flora?

Destructive as they may be, extinctions also have a positive side, giving new species the chance to evolve and occupy the empty ecological niches. It is believed that mammals took advantage of the extinction of the dinosaurs to expand rapidly ... and here we are. But if such catastrophic events happen too often, life might not be able to maintain itself. Fortunately, we on Earth may benefit from a reliable sentry, Jupiter.

With a mass 300 times that of Earth, one theory suggests that Jupiter actually intercepts most of the objects that might threaten us. Some estimates suggest that, in the absence of our giant guardian, the Earth would be hit by a 10 km-size body every 10 000 years instead of every 100 million years [43]. At such a high impact rate, it is doubtful that life could ever have developed on our planet.

67 Where did the asteroid implicated in the extinction of the dinosaurs fall?

The extinction of species associated with the transition between the Cretaceous and the Tertiary periods (referred to as the *K-T* transition) 65 million years ago was the object of several studies in the 1970s. It was a massive event: all dinosaurs disappeared along with every land animal over 25 kg. While studying the corresponding marine sedimentary layers in several parts of the world (Italy, Denmark, and New Zealand), the American geologist, Walter Alvarez, found that they all had an unusually high iridium content. The element iridium is rare in the terrestrial crust because most of it sank to the core, along

with iron and the other heavy elements, during the differentiation phase of our planet's formation (Q. 52). Any iridium now found in the Earth's rocky crust originates from "cosmic dust," i.e. the meteor dust that continuously rains down on Earth. In 1980 Walter Alvarez and two colleagues, together with Walter's father, Physics Nobel Prize Luis Alvarez, published a sensational article claiming that the high content of iridium in the *K-T* layer was due to our collision with an asteroid or a comet 10 kilometers in diameter [1]. Since then, iridium has been detected everywhere the corresponding sedimentary layer is found, confirming their theory.

A certain large crater in the Yucatan peninsula may well be due to this asteroid. Debris from the explosion has been identified up to 2000 km away. A tsunami with a 200 m-high wave would have swept the whole Gulf of Mexico. Enormous quantities of dust would have been projected into the stratosphere, covering the entire globe, eclipsing

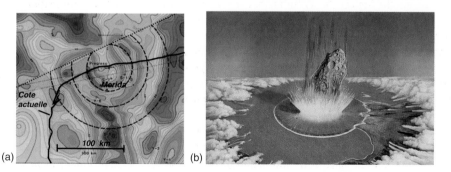

(a) Map of the gravity anomaly showing the size of the 180 km diameter crater in Yucatan, and (b) an artist's view of an asteroid impact on Earth. Credit: LPI and D. Norton.

sunlight, and stifling photosynthesis for decades. As the larger debris objects fell back through the atmosphere, they would have been heated to incandescent temperatures and triggered immense firestorms all over the continents, increasing atmospheric pollution and destroying the shelters and food supplies of most animals [28].

68 What could be done if an asteroid threatened to collide with Earth?

Several methods have been proposed. A rocket might be sent out to attach itself to the asteroid and change its course. Anchoring it to the asteroid would be a delicate operation, however, and if the asteroid was rotating on its axis, the action of the rocket would be ineffective. In another proposed method, instead of the rocket giving the asteroid a gentle push, it could ram into it hard, like a billiard ball. The risk in this type of operation is that the asteroid might break up into large fragments that would continue on their trajectory and still cause havoc when they collided with us.

Credit: D. Durda, FIAAA/B612 Foundation.

Another solution involves exploding a bomb on the asteroid to pulverize it, but again, much of the debris would continue towards the Earth. A better solution recently proposed by two former astronauts [30] is to deflect the object using gravity. A special 20 ton spacecraft would be sent *close* to the asteroid (see image) and exercise a weak gravitational attraction on it; after a few months, it would actually deflect the threatening object.

69 What is the Kuiper Belt?

The Kuiper Belt is a region of the solar system disk (i.e. in the plane of the planets' orbits) that extends from the orbit of Neptune (at 30 AU from the Sun) to approximately 50 AU from the Sun. It contains mainly small bodies 50–1000 km across called *KBOs*, for Kuiper Belt objects. Pluto (diameter 2400 km) is thought to be an object originally from the Kuiper Belt, as well as Eris (diameter 2600 km). Both now orbit outside the belt and are believed to have been ejected from it under the influence of Neptune.

The Kuiper Belt is similar to the main asteroid belt except that its objects have higher ice contents (like comets) and are not as rocky as asteroids. The belt is also larger and contains a greater number of objects than the main asteroid belt; its total mass is estimated to be about 1/10th that of Earth. It is named after the Dutch-American astronomer Gerard Kuiper (1905–73) who predicted in the 1950s that certain comets originated in that region. The first KBO was discovered in 1992.

The many *KBOs* (green dots), small scattered bodies (brown dots), the Trojan satellites (in violet) of Jupiter and Neptune, respectively, and the Sun (yellow circle) (Q. 63).

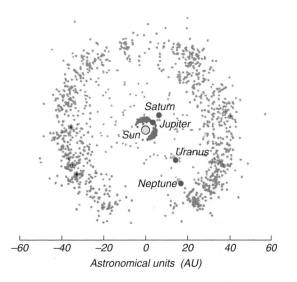

The Kuiper Belt is a remnant of the young solar system's disk of planetesimals. Left "unused," out beyond the orbit of Jupiter, they cluster in this band under the gravitational pull of the giant planets (Jupiter, Saturn, and Uranus).

70 Where do comets come from?

Over 3000 comets have already been observed and a new one is discovered about once a month. Comets belong to one of two families, depending on their period (the time it takes for them to go around the Sun): short (less than 200 years) or long (from 200 to several tens of thousands of years). The origins of the two types are different.

Short-period comets originate in the Kuiper Belt: Neptune pulls them out of the belt as it passes through. Long-period comets, which have extremely elliptical orbits, come from a vast reservoir of icy bodies that extends in all directions as far out as 100 000 AU, half the distance to the nearest star (Proxima Centauri). This region, which has never been directly observed because it is too distant, is named the Oort Cloud after the Dutch astronomer, Jan Oort, who proposed its existence in 1950.

This putative cloud of objects is thought to be the remains of the swarm of primordial planetesimals out of which the giant planets (Jupiter, Saturn, Uranus) formed. Myriads of these small objects would have been ejected into highly elongated orbits by the giant planets in much the same way that spacecraft are flung away by *gravity assist* (Q. 78). Then, perturbations from nearby stars would have

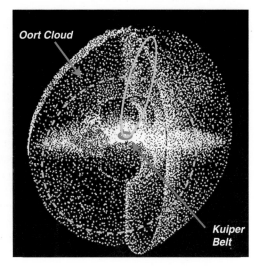

tilted their planes in all directions and made their orbit more circular. This would explain why the Oort Cloud is spherical in shape, unlike the Kuiper Belt which is well settled into the orbital plane of the Solar System. Most of the trillions of Oort Cloud bodies, being weakly bound to the Sun, should have more or less circular orbits (yellow broken line on figure). Rare indeed would be those that fall into the inner solar system (yellow solid line) to become the comets that we see, kicked out of their orbits by the gravitational effects of nearby stars or even by tidal effects generated by the Milky Way.

71 How big are comets?

Halley's comet imaged by the European Giotto spacecraft in 1986. The nucleus, with its jets of dust illuminated by the Sun, is approximately 15 km long. Credit: ESA.

Comets are small objects about 10 km across. A mix of dust, rock, and "ices" of water, methane, ammonia, and carbon dioxide, they have been described as "dirty snowballs." The nucleus of a comet is too small to be resolved with the naked eye or even with our ground-based telescopes. What we do see is the *coma*, an envelope of gas evaporating from the nucleus as the comet approaches the Sun. The coma shines by reflected solar light but also emits some phosphorescent light. It is about one million km in diameter. The tails of comets are much longer than that, in some cases as long as the distance from Earth to the Sun.

72 What is a comet's tail made of?

Comet Hale Bopp in 1997. Credit: Bengt Ask.

Comets normally have not one tail, but two, a white one and a blue one. The white tail is caused by dust particles evaporating from the comet's nucleus due to solar heating and radiation pressure (Q. 46). The dust reflects sunlight and the tail looks white.

The blue tail is caused by the solar wind (Q. 47); the protons of the "wind" strike the atoms and molecules of the comet's nucleus and ionize them (knock out some of their electrons). This tail is self-luminous, as the ionized material fluoresces.

73 In the age of space probes, is it still useful to observe the planets with telescopes?

Direct planetary exploration with spacecraft has produced a formidable harvest of images and scientific data that we never would have obtained with ground-based or orbiting telescopes. Nevertheless, telescopes continue to play an important role.

Space probes can observe planets and their satellites only during the brief period of their flight or lifetime, while telescopes can observe them repeatedly and detect rare events or monitor variable or short-lived phenomena. Moreover, spacecraft do not see everything. The Hubble Space Telescope recently discovered two rings around Uranus that Voyager 2 had missed during its passage in 1986.

(a)   (b)

Two examples of telescope contributions to the study of planets: (a) a polar aurora on Jupiter and (b) a giant dust storm on Mars. Both photos by the Hubble Space Telescope.

74 What do the Mars rovers do?

First, there was Sojourner in 1997, and then the twin rovers *Spirit* and *Opportunity* which began exploring Mars in 2004 and delighted the scientific community with their productivity and longevity. Soon, there will be *Mars Science Laboratory*, a highly sophisticated rover scheduled to arrive in 2012. The purpose of all these rovers is to study the geology of the red planet and try to determine if life has ever been present there.

The current rover twins have been studying the surface of Mars as geologists would. Each is equipped with two cameras mounted on a 1.5 m mast that provides

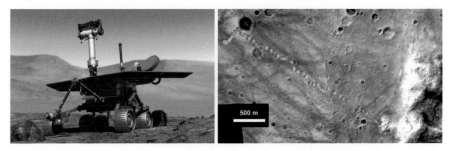

(a) The Mars robot *Spirit*, and (b) its track on the Martian surface photographed in 2004 by the orbiting Mars Global Surveyor. Credit: NASA.

it with stereoscopic viewing, and an articulated arm capable of positioning measuring instruments on the ground. Their artificial hands have built-in microscopes and abrasive tools that work like a geologist's magnifying glass and hammer to investigate what lies under the weathered outer layer of Martian rocks.

75 Why colonize Mars?

Mars, the red planet, the only other planet in the solar system that might be habitable, has excited the imagination for centuries. Mars is smaller than Earth, but its surface area is equal to that of our continents. It has seasons. There is water in the polar caps and underground. Its atmosphere (95%) carbon dioxide is low in density, but provides at least some protection against solar and cosmic radiation. It also offers crucial assistance in slowing down spacecraft during landing operations. The length of the Martian day (24 h 39 min) is similar to ours. It is cold on Mars, but the average temperature, $-63\,°C$, is not that much lower than in northern Siberia or the Canadian arctic regions in winter.

A trip to Mars takes about nine months, which is only about two to three times as long as the trip made by nineteenth century Europeans emigrating to Australia. The colonization of Mars is bound to happen one day.

Until that day the exploration of Mars, first with robots, later with human explorers, will remain of great scientific interest: in furthering our understanding of its geology,

Artist's view of human explorers on Mars. Credit: NASA.

particularly its volcanic activity and the role that water played in its past; in contributing to our understanding of the evolution of the solar system and, through that, of the Earth; and in answering our questions about the possibility of Mars harboring life in the past and the eventuality of an exchange of living material at some time between Mars and the Earth.

76 Which way to Mars?

You cannot get to Mars by aiming straight for it, as though you were shooting at a target because, in this case, both the gun and the target are moving. To go from one planet to another, the most economical path from the point of view of energy is half of an elliptical trajectory that requires only two firings, one to leave the orbit of the planet you are departing, and the other, at the end of the trip, to inject you into the orbit of the planet you are visiting.

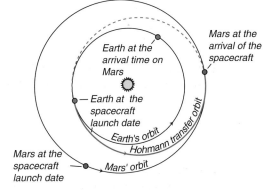

It was the German engineer Walter Hohmann (1880–1945) who established this astronautical concept in 1925, at a time when interplanetary travel was still a dream. The trajectory is called the *Hohmann transfer orbit*.

You cannot take off for Mars just any time you choose, either. You must work with the schedule. Launch dates have to be timed in such a way as to guarantee that, when the rocket arrives at the orbit of its destination planet, the planet is actually there. In the case of Earth and Mars, the proper alignment between the two planets, called the "planetary launch window," recurs once every 26 months. For the return part of the trip, the correct configuration is again necessary, and this only appears again 15 months later. Since the trip in each direction requires about 8.5 months, a round trip for a Mars mission requires a minimum of 32 months, or 2 years and 8 months.

Walter Hohmann.

77 What is solar sailing?

A solar sail does not, as one might expect, make use of the *solar wind*, the stream of protons and electrons emitted by the Sun; it uses the *pressure* of solar radiation, i.e. the pressure of the light emitted by the Sun in the form of photons (Q. 46). The particles making up the solar wind are harmful, but the pressure they exert on a surface is completely negligible compared to the solar radiation pressure, which itself is minute. In empty space, however, where there is no friction to slow down movement, a sail with an area of a few hundred square meters could carry a useful load of several hundred kilograms, and this without using any fuel or having any limitations on its range of travel.

The pressure from solar radiation acting on a sail is the sum of the force supplied by the bombarding photons and the force exerted by the kickback as they are reflected. For a highly reflective sail, the total force acts in a direction roughly perpendicular to the sail. How can we exploit this? By orienting the sail in such a way that this combined force slows or accelerates the orbital motion of the craft, as desired. If the force is applied in the direction opposite to that of the orbital movement (case (a) in the figure), the solar craft will slow down and move nearer to the Sun; if applied in the direction of motion (case (b)), the craft will accelerate and move away from the Sun.

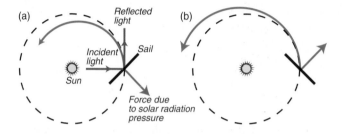

A solar craft cannot take off directly from the ground. Before its sail is deployed it must be boosted into an orbit outside the Earth's atmosphere by means of an ordinary rocket. The acceleration of a solar craft depends on the surface area of its sails, but is always quite small. On the other hand, it is continuous, unlike rocket engines that work for only a few minutes. With a sail, the travel time to Mars would be about 18 months, and while solar radiation diminishes as the distance from the Sun increases, the acceleration acquired when closer to the Sun is sufficient to plan for travel to the asteroid belt, to the giant planets, and beyond.

NASA is currently studying the *Interstellar Probe*, a solar sail spacecraft that could travel to the periphery of the solar system in 8 years, compared to the 40 years required by *Voyager I*.

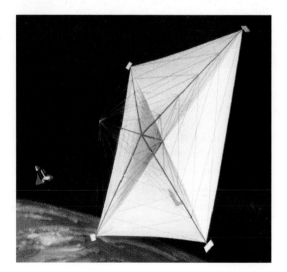

Artist's view of a solar sail. Credit: NASA.

78 How could the Voyagers explore so many planets and satellites in one trip?

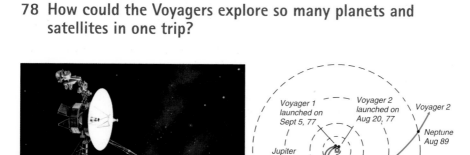

The two NASA interplanetary probes, *Voyager 1* and *2*, are the Christopher Columbuses and Captain Cooks of our age. These amazing machines have explored the solar system beyond Mars, including the giant planets Jupiter, Saturn, Uranus, Neptune, and 48 of their satellites. Their extraordinary harvest of images, spectroscopic measurements, magnetic field mappings, and cosmic ray characterizations has revolutionized planetary studies. They have gone further into outer space than any other man-made object, and are now more than 100 AU (15 billion km) from Earth. They recently reached the very frontiers of the solar system (Q. 35) and are continuing on outward.

Launched in 1977, the two Voyagers were designed to exploit the chance alignment of the four giant planets that would permit them to visit them all, in record time, and using the minimum fuel. In designing such missions, the trick is to carefully plan the trajectory of a spacecraft and its approach to a planet so as to take advantage of that planet's gravity to boost the acceleration of a probe, similar to the way a cyclist uses the speed acquired while going down a hill to start up the next one. In space travel, this type of boost is called a "gravity assist."

Too far from the Sun to be able to count on solar panels, the probes are equipped with small thermoelectric nuclear reactors which produce the 300 W they require. To communicate with them over the vast distance separating them from Earth, NASA uses radio antennas 70 m in diameter. After more than 30 years, these marvelous little "spacecraft that could," are still at it!

The Earth

79 How was the size of the Earth measured?

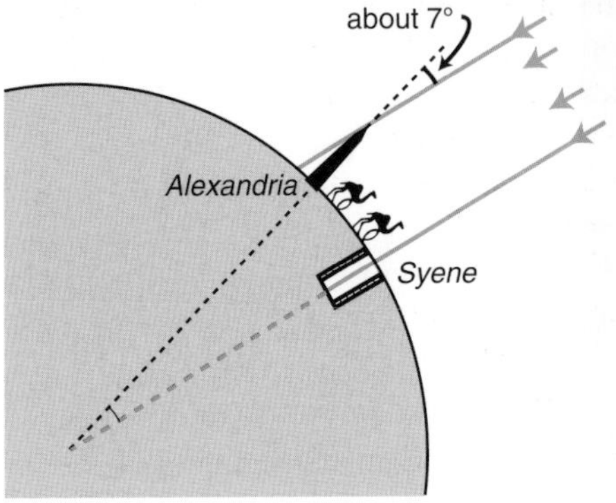

about 7°

Alexandria

Syene

The earliest measurement of the size of the Earth was made by the Greek astronomer Eratosthenes, director of the great library of Alexandria (Q. 192), in about 240 BC. No Greek astronomer doubted the sphericity of the Earth back then,[†] and learning that a ray of sunlight penetrated to the bottom of a deep well in Syene (today, the city of Aswan), south of Alexandria, on the day the Sun appeared the highest in the sky (i.e. the summer solstice), gave him the opportunity to determine the radius of the Earth. He measured the zenithal angle of the Sun in Alexandria at the solstice and found it to be 1/50th of a circle (about 7°).

Assuming that the Sun was very far away,[‡] and that its rays were parallel, Eratosthenes concluded that the Earth's circumference was 50 times the distance between Alexandria and Syene, estimated by travelers to be 5000 stadia. From that he deduced a circumference of 250 000 stadia for the Earth. There is some doubt about the exact value of the stadion unit used by Eratosthenes, but the error in his measurement is probably less than 10%, a real feat for the time. This measurement of the size of the Earth is one of the greatest scientific achievements of antiquity.

It was not until 1670 that a French astronomer, Jean Picard, made a more accurate measurement. Picard used the same general method as Eratosthenes – measuring the length of an arc of meridian – this time between Paris and Amiens, a city about 1° latitude north of Paris. The latitudes of the two cities were easily obtained by celestial sighting; the difficulty was in precisely measuring their distance. Picard used the newly developed method of geodesic triangulation, measuring angles between benchmarks (hills, towers, steeples) forming the apexes of a chain of 13 triangles. By measuring one specific side (the *base*) of one of the triangles, the chain of triangles could be fully determined, hence its length (e.g. length *A* to *D* in the figure). The base Picard used

---

[†] Unfortunately, Saint Augustine (354–430 AD) found it absurd to think that there might be descendents of Adam at the antipodes (*anti*, opposed and *podes*, feet), with their feet "opposite" to his own, and he declared that the Earth was flat – a reversal of knowledge that was to be corrected only in the late Middle Ages.

[‡] A few centuries later the Chinese made the opposite assumption, that the Sun was fairly close, and, also convinced that the Earth was flat, derived a completely erroneous distance to the Sun from a similar angular measurement.

was a straight stretch of road about 11 km long. The value he derived for the radius of Earth was remarkably accurate, with an error smaller than 0.5%.

The measurement was repeated with a longer base late in the eighteenth century by French astronomers Joseph Delambre and Pierre Méchain, using the arc of meridian between Dunkirk in Northern France and Barcelona, Spain. Their real goal at that time was to define the *meter*, the standard unit for the metric decimal system. In 1790, the new French revolutionary government had decided to establish standards for weight, volume, and length that would apply throughout the land, and imbued with the spirit of universality born of the Revolution, they hoped that the new system would be adopted by other nations. So the standards, they felt, should be "neutral," without French connotations, and what could be more universal than a unit of length based on the size of the Earth? It would belong to all humankind! The meter, it was decided, would be defined as the 1/10-millionth of a quarter of the Earth's circumference. Unfortunately, a small error crept into their measurements, and the meter is in fact 0.2 mm shorter than it should have been.

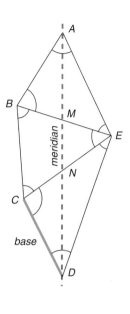

Since then, a variety of methods have been employed to determine the shape and size of the Earth, including some involving spacecraft. As it turns out, our planet is not a perfect sphere: it is flattened at the poles and swollen at the equator, approximating an ellipsoid with a short axis (the polar radius) of 6 356.750 km and a long axis (the equatorial radius) of 6 378.135 km. A more accurate shape to describe it is the special surface called a *geoid*. This is defined as the surface around the Earth that is everywhere perpendicular to the direction of gravity and coincides with mean sea level. The geoid, which deviates from the ellipsoid by −106 m to +85 m, depending on location, serves as a reference for determining altitudes on Earth.

80 How was the mass of the Earth measured?

Determining the mass of the Earth is not an easy task. Isaac Newton had to come onto the scene with his law of gravitation, published in 1687, before it could be undertaken (Q. 194).

Starting with Kepler's law of planetary motion, Newton demonstrated that two objects of mass m and M are attracted by a force, F, which is proportional to their masses and inversely proportional to the square of their distance d: $F = GmM/d^2$, where G is a constant called the *gravitational constant*. He also showed that the gravitational attraction on an object at any distance from a spherical body is the same as if all the mass of that body were concentrated at its center. At the surface of the Earth, the gravitational attraction on a body of mass m (i.e. its *weight*) is equal to mg, where g is the acceleration due to gravity (Newton's second law). The force of weight is simply the gravitational

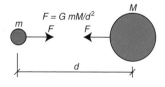

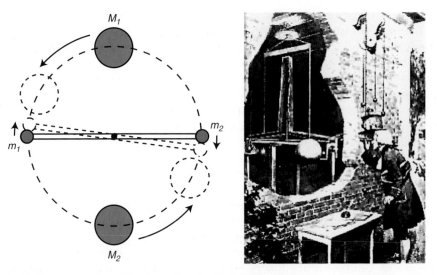

In his famous experiment for "weighing the Earth," Cavendish measured the rotation of a rod attached to two small balls (m_1 and m_2) while he moved two large balls, M_1 and M_2, close to them.

attraction of the Earth. And if R is the radius of Earth (assumed to be spherical), we have $mg = mMG/R^2$, where M is the mass of the Earth. Hence $M = gR^2/G$.

The acceleration of gravity, g, is easy to measure, using, for example, a pendulum whose period of oscillation is a function of its length and of g. We know the radius of the Earth, R. To calculate the mass of the Earth, we just need to find the value of G.

It was the brilliant but taciturn British physicist Henry Cavendish[†] who, in 1798, made the first successful measurement. The apparatus he used was extremely simple: two small lead spheres suspended at the ends of a wooden rod hung from a wire, and two large lead spheres.

When he moved the two large spheres, M_1 and M_2, close to the two small ones, m_1 and m_2, the small ones were attracted by the large ones, twisting the wire from which they were suspended. Having calibrated the wire's resistance to torsion and measured the angle through which it turned, Cavendish derived the force of attraction between the large and small balls, hence the value of G.

The difficulty of the experiment lay in the fact that the attractive force between the lead spheres is minute and the measurements easily perturbed by differences in temperature and air currents. Cavendish had enclosed his apparatus in an isolated room while he, himself, stayed outside, moving the large spheres from a distance and observing the movements of the small lead balls through a telescope.

[†] Henry Cavendish, although an extremely rich man, lived as a recluse. It was said that he spoke no more in his whole life than a Trappist monk. However, he was one of the greatest physicists and chemists of his time: he determined the compositions of air and water, discovered hydrogen, and conducted the earliest studies of electricity.

Cavendish's experiment, dubbed the "scale that weighed the Earth," remains justly famous. It is one of the finest and most important scientific experiments in history. There was something almost magical about it. How could four balls inside a brick shelter tell us the mass of the Earth? But the magic did not stop there: once the gravitational constant was known, it could even tell us the mass of the Sun, the planets, and the entire Universe!

The accuracy of Cavendish's measurement, excellent for his time (7%), has been improved on only recently. Today's value for G is $6.674 \cdot 10^{-11}$ $m^3 kg^{-1} s^{-2}$. The derived mass M for the Earth is about $5.6 \cdot 10^{24}$ kg, and its mean density 5.5.

81 How old is the Earth?

For most of human history, pronouncements concerning the age of the Earth were based on religious dogma and metaphysical speculation. The announced values ran from only a few thousands years – based on counting the successive generations of Adam's descendants listed in Biblical texts – to an infinite length of time for Aristotle, who thought that the Earth had always existed, or again to the eternal cycles of Oriental mythologies.

The first scientific estimates came in the eighteenth century, thanks to the burgeoning of geology and theoretical approaches by physicists. Geologists deduced ages of several million years based on the time required to accumulate the enormous thicknesses of Earth's sedimentary rock layers and account for the slow erosion of mountains and other geological formations. Simultaneously, however, the French naturalist, the Count of Buffon (1707–88), proposed that the Earth had originated in a plume of intensely hot material torn from the Sun that had then cooled and solidified. The experiments he conducted at ironworks, heating spheres of metals, clay, and rock, then measuring their cooling times, led him to deduce that 76 000 years must have elapsed for the Earth to reach is current temperature.[†]

This opposition between the geologists working empirically and the physicists trusting too much in their theoretical laws lasted until early in the twentieth century. It reached its climax in 1862 with the debate between, on one side, the Scottish physicist

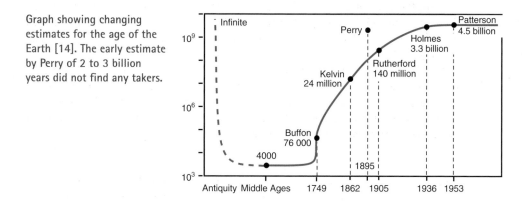

Graph showing changing estimates for the age of the Earth [14]. The early estimate by Perry of 2 to 3 billion years did not find any takers.

[†] The book in which Buffon proposed this age was condemned by the Vatican.

William Thomson and, on the other, geologists allied with the supporters of Darwin's theory of evolution.

Following in the footsteps of Buffon, Thomson (later to become Lord Kelvin) hypothesized that the young Earth was originally at the temperature of molten rock (about 4000 °C) and, applying the laws of heat diffusion by conduction, he calculated that no more than 96 million years were needed for Earth's temperature to drop to present levels,[†] though he actually proposed a range of ages between 24 and 400 million years to allow for the uncertainties in his hypothesis. For the Sun, whose shining he assumed was due to its gravitational contraction (Q. 33), he derived an age of between 20 and 25 million years. That then set an upper limit for the age of the Earth, which could not have been born before the Sun. The Earth was probably 24 million years old, he concluded.

Geologists and biologists, meanwhile, were estimating that even 400 million years were not enough to account for the slow processes of sedimentation and natural selection that they were observing. Thomson's reputation for scientific rigor was so powerful that it was difficult to contradict him, however, and astronomers also rallied to his views. Hence, for almost half a century, the age of the Earth was locked in at 24 million years, to the despair of geologists. But Thomson was wrong, as one of his assistants, John Perry, showed in 1895. Thomson's error had been to assume that the interior of the Earth was motionless, which is not the case, at least over time lapses of millions of years. The mantle is actually viscous and is host to large movements that transport heat from the interior outward, a phenomenon called *convection*. This means that the core of the Earth did cool down faster than Thomson had inferred, and that the heat released in the process contributes in large part to the surface heat flow measured. As a result, the age calculated by Thomson was much too low. Perry derived an age of 2 to 3 billion years [34]. Unfortunately, his work was ignored and Thomson's estimate persisted.

A few years later, French physicist Henri Becquerel discovered radioactivity, and his colleagues Pierre and Marie Curie showed that it produced heat. Once it was known that the Earth's crust contains radioactive rocks that are a significant source of heat, the old

William Thomson (1824–1907), the future Lord Kelvin.

disagreement between geologists and physicists melted away. But, irony of ironies, we now know that the contribution of heat from radioactivity was seriously overestimated then, and that Perry's analysis was the correct one. All the same, Thomson's theoretical approach – as corrected by Perry – was right.

Can the age of the Earth be determined directly? Radioactivity actually provides a powerful tool to derive the age of the terrestrial crust because the abundance of the different radioactive elements is a function of time and can be used as a clock. This method of *radiometric dating* came to be employed very quickly. In 1905, Rutherford calculated an age of 140 million years for Earth, and many other measurements soon

---

[†] This flux is 80 mW/m², corresponding to a temperature gradient in the Earth's crust of 36 °C/km.

(a) Clair Patterson (1922–95) determined the age of the Earth from a meteorite found in Arizona (b).

followed his. Finally, in 1936, Holmes showed that the oldest rocks at the centers of continents were 3.3 billion years old, and even older rocks have been found since then, particularly in Canada, where some over 4 billion years have been identified. Unfortunately, radiometric dating cannot take us beyond the time when the rocks began solidifying or were transformed. To go back beyond that, astronomy must be invoked, and theories about the formation of the solar system.

The Earth and the planets formed by the accretion of dust grains in the protoplanetary disk (Q. 33). Most of the meteorites that fall on Earth date from this same ancient time. Although some of them originate from recent collisions of asteroids or are impact debris from neighboring planets, the oldest are from the distant age of the Earth's formation. In 1953, American geochemist Clair Patterson analyzed a meteorite that had fallen in Arizona about 40 000 years ago and found it to be 4.55 billion years old. This is our best current estimate for the age of the Earth.

82 What is inside the Earth?

A century and a half after Jules Verne's visionary "Voyage to the Center of the Earth," the interior of our planet remains mysterious and essentially unexplored. The deepest hole ever drilled has penetrated only 12 km. At that depth, conditions are already extreme: the pressure is 4000 times the atmospheric pressure and the temperature 500 °C. At present, it is technically impossible to drill deeper.

Nevertheless, we have been able to piece together a good portrait of the Earth's interior thanks to the many seismic studies being carried out around the world. Numerous, permanently installed seismographs monitor the thousands of earthquakes that take place each year, showing the propagation of seismic waves in the interior of the Earth and allowing us to derive its structure, in the manner of sonograms that let us discern a fetus in its mother's womb.

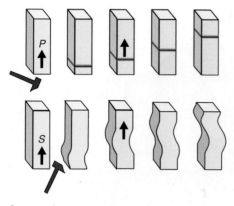

Compression waves, (P), generated when a beam is struck at one end, and transverse waves, (S), when struck on its side.

Seismic waves propagate both along the surface of the Earth and through its interior. The interior waves are of two types, P and S, for primary and secondary, so named because once generated, the P-wave is faster and arrives at a distant detection site first, and the S-wave comes in second. P-waves are compression waves, like sound waves, while S-waves are transverse (or shear) waves, like waves on the surface of water. The velocity of both types is a function of the material they pass through. Like light, seismic waves are refracted (i.e. change direction) when passing from one medium to another and are also reflected at the corresponding boundaries. The S-waves do not propagate in liquids.

When measurements from the global network of seismographs were analyzed, it was learned that there are zones where the S- or P-waves did not arrive. From that, it was deduced that the outer shell of the core was liquid. Surrounding that shell is the Earth's *mantle*, which is plastic but acts as a solid insofar as the P- and S-waves are concerned. It was also learned that the central part of the core is solid. This was established by observing the refraction of the P-waves passing through it and their reflection at the transition between the solid and liquid zones. That finding was predictable – the central part of the core could only be in a solid state because of the enormous pressure prevailing there.

What is the composition of the Earth's interior? The materials of the crust and upper mantle are accessible by direct sampling or during volcanic eruptions. The crust, which is only 5 km thick under the oceans and about 50 km thick under the continents, is composed of granite for the continental part and basalts for the oceanic floors. As for the upper mantle, it is about 400 km thick and is essentially made up of silicates of iron and magnesium.

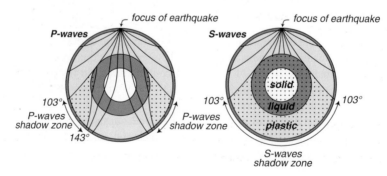

The "shadow" zones associated with S-waves have allowed us to map the liquid parts of Earth's interior (the trajectory of P-waves is more complicated than represented here).

As far as the deep mantle and core are concerned, we have to rely on theory. The Earth formed by the accretion of interstellar dust, and meteorites represent samples of that ancient material. Accretion, gravitational forces, and radioactive heating caused those materials to differentiate (Q. 52): the densest ones, such as iron, sank into the interior, while the lighter ones, the silicates in particular, floated to the surface. The current understanding is thus that the planet is composed of an external envelope made up mainly of aluminum silicates (the crust) floating on a viscous layer of silicates rich in iron and magnesium (the mantle). These layers surround the Earth's central core, which is composed of iron and some nickel.

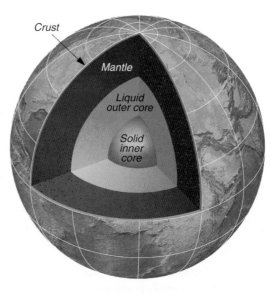

83 Where did the water on Earth come from?

A photo taken from space shows something striking about Earth: our planet is blue. Oceans occupy 71% of the Earth's surface and their volume is enormous: 1.4 billion cubic km. Their average depth is 3800 m, five times the average height of the continents when measured from sea level.[†] Where did all this water come from?

A molecule of water is composed of two atoms of hydrogen combined with one of oxygen. Hydrogen, the lightest chemical element, is found throughout the Universe. It is the fundamental substance that makes up the stars. Oxygen atoms, which are much heavier than hydrogen, are synthesized in novae and supernovae (Q. 16). Our solar system formed from a cloud of "stardust" composed of the debris from such a stellar explosion (Q. 33). The young Sun heated the cloud and, when it was warm enough for the reaction to take place, the hydrogen and oxygen atoms combined to form water.

Water atoms were thus present in the disk of debris left orbiting the Sun after its formation, and the planets, including Earth, were formed

View of Earth from space. Credit: NASA.

[†] The oceans contain most of Earth's water: 97.2%. As for the rest, 2.1% is bound up in glaciers, and only 0.7% is in rivers, lakes, and aquifers. The atmosphere holds only 0.001%.

from this leftover matter. How did our planet manage to capture so much water? There are three current theories.

According to the first, water in the form of ice particles was simply one constituent of the original disk orbiting the young Sun, and this material accreted to form the Earth. Then, as gravity compressed and heated the accreted particles, the water was vaporized and evacuated by a myriad of volcanoes. Our primitive atmosphere was the result. Five hundred million years later the temperature of this first water vapor atmosphere dropped below the boiling point and the vapor condensed to liquid water and fell to the surface in a deluge of warm rain. The water ran down to the lowest levels of the young terrestrial crust, accumulated into puddles, then lakes, and finally filled the seas and oceans.

According to the second theory, the water came from the massive bombardment of dust and ice fragments (left over after formation of the planets) to which Earth was subjected during its first billion years of existence.

In the third theory, the water was brought to Earth by the comets that collided with it early in its history (Q. 70). Comets, which are also essentially clumps of debris left over from the formation of the solar system, are rich in water ice.

Although it is not yet clear which of these theories will prove to be the correct one, it is likely that all three sources contributed to the formation of our oceans. Volcanic emissions seem to have been the biggest contributors, however.

84 Do any of the other planets have oceans?

Of all the planets in our solar system, only on Earth does water fall from the sky as rain, only on Earth are there rivers and oceans of life-sustaining water. The giant planets (Jupiter, Saturn, Uranus, and Neptune) have only very small solid cores surrounded by liquefied gas under intense pressure. No chance of finding continents or oceans there! The inner planets (Mars, Venus, Mercury) are rocky, more like Earth, but only Earth is lucky enough to have reserves of liquid water.

Mercury has hardly any atmosphere and temperatures on the side facing the Sun are high enough to melt lead (420 °C).

Venus, our sister planet, is similar to Earth in size and mass, which once led some to believe that it was cloaked in vegetation, even inhabited. But 40 years of exploration by space probes has made it clear that Venus is a lifeless desert. Being closer to the Sun than we are and also subjected to a devastating greenhouse effect, any liquid water once present there evaporated long ago in the hellish temperatures that now reign at its surface (460 °C).

Mars, being further from the Sun than Earth, is colder than our planet. In the nineteenth century certain marks on its surface made some observers think that they were seeing canals, but we know now that this was an illusion. Nevertheless, space missions have revealed what look to be dry riverbeds on Mars, indicating that there was indeed running water on the surface early in the planet's history. What remains of that water is now buried in the Martian soil in the form of small amounts of water ice.

A recent surprise is that, among the satellites of the outer planets, it is likely that Europa (a moon of Jupiter) has a liquid ocean of water covered by a thick layer of ice, and that Titan (one of Saturn's moons) may have seas or lakes of ethane or methane.

85 Where does the oxygen of our atmosphere come from?

The primordial solar nebula that produced Earth (Q. 33) was essentially composed of hydrogen and helium gas mixed with fine dust. There was no free oxygen gas in the cloud because this extremely reactive element had already combined with the other chemical elements in the dust particles, iron and silicon in particular.

The original hydrogen and helium in the atmosphere of the young Earth were lost to space under the effects of rising temperatures (Q. 51) and the solar wind. Around 4.4 billion years ago, when the Earth had cooled down enough for a crust to form, intense volcanic activity began producing huge quantities of carbon dioxide, ammonia, and water vapor. The carbon dioxide of this primitive atmosphere triggered a greenhouse effect that prevented the Earth from cooling too much (Q. 92), but there was still no free oxygen.

As temperatures dropped, the water vapor condensed to form oceans (Q. 83). The carbon dioxide content of the atmosphere was reduced, then, as that gas dissolved in the young oceans, and precipitated out as carbonates.

Life began in shallow water about 3.5 billion years ago with organisms similar to bacteria, the *prokaryotes*, that did not produce oxygen. Then, about 2.7 billion years ago, a crucial event took place: cyanobacteria appeared. These tiny, single-celled entities used photosynthesis to convert carbon dioxide into organic carbon and oxygen gas.[†] At first, this new free oxygen was completely absorbed by the iron present in reduced form in the water and by the dissolved hydrogen gas produced by volcanoes, but then, around 2.4 billion years ago, its concentration in the atmosphere shot up. Cyanobacteria had no predators back then and were able to proliferate unchecked, turning the seas into giant factories of oxygen production. The results were almost instantaneous on geological timescales: a brutal enrichment in oxygen called the "oxygen catastrophe."

To complete the picture, algae, which appeared 2.1 billions years ago, followed by terrestrial plants, 500 million years ago, subsequently added their own photosynthesis

Fossil cyanobacteria, the type of photosynthesizing bacteria responsible for producing our atmospheric oxygen. These are about 0.01 mm in size.

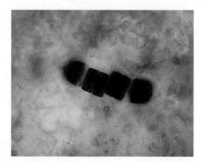

[†] It is chlorophyll, the "green molecule," that starts the process, absorbing solar energy and transferring it to other molecules in a cascade of chemical reactions. The details are complex but can be summarized as follows: solar energy + carbon dioxide + water → glucose + oxygen. In chemical notation: photons + 6 CO_2 + 6 H_2O → $C_6H_{12}O_6$ + 6 O_2

activity to boost the supply of oxygen in the atmosphere to the levels that we have today.

86 What causes the seasons?

It is a widespread misconception that summer is the warmest season because the Earth is closer to the Sun then. Although it is true that our orbit is slightly elliptical, there is very little difference in our distance to the Sun. In fact, we are closest to it in January. We have seasons because the axis of rotation of the Earth is inclined with respect to its orbital plane. As the Earth revolves around the Sun, the direction of its rotation axis remains essentially fixed in space (pointing towards the star *Polaris*), but its inclination *with respect to the Sun* changes over the year and causes the seasons.

When our inclination towards the Sun is greatest, we experience a *solstice*. The term means "sun stopping" in Latin and refers to the fact that, during a solstice, the height of the Sun at noon appears to be constant over a period of several days. When the inclination is zero, the event is called an *equinox*, meaning "equal night." During equinoxes, nights and days are of equal length (12 h) everywhere on Earth. Due to variations in the calendar, leap years in particular, the dates for solstices and equinoxes vary by as much as three days from year to year.

At the summer solstice for the northern hemisphere, the North Pole is inclined towards the Sun, the Sun's rays are the least oblique as they strike the Earth, and a given area on Earth receives the most energy. At the winter solstice, the opposite is true, the rays are the most oblique and a given area on Earth receives the least energy. It is important to remember that seasons are reversed from one hemisphere to the other. For example, when it is winter in the northern hemisphere, it is summer in the southern hemisphere – and vice versa.

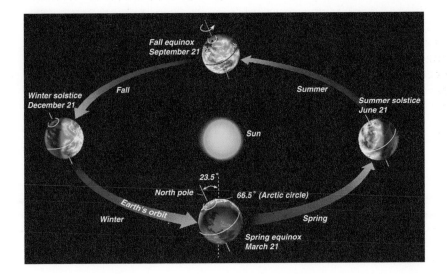

However, it is not at the winter solstice that the coldest winter temperatures are reached. That happens a few weeks later. Let us assume the case of the northern hemisphere for a moment as we examine why. The winter solstice occurs there around December 21, but the Earth's crust in that hemisphere has conserved a good

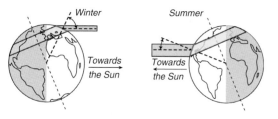

deal of heat from the previous summer, and it will be several more weeks – late January or early February – before it cools down. The same reasoning applies to summer temperatures in reverse: they peak in late July rather than at the summer solstice around June 21.

This temperature lag is the basis for the western tradition of defining the start of each season as coinciding with a solstice or equinox. Other calendars do not take the lag into account. In China, for example, solstices and equinoxes mark the approximate middle of a season: the Chinese summer begins at about May 7, fall at around August 6, winter at about November 8, and spring (which starts the Chinese New Year) at around February 5.

Because of its elliptical orbit, the Earth's speed varies slightly during the year and the seasons are not all equal in length. The Earth travels faster on its orbit when closer to the Sun. For the northern hemisphere, this happens in winter, and winter is thus the shortest season. Spring, summer, fall, and winter, respectively, last about 92, 93, 90, and 89 days – differences that Greek astronomers had already noted in the fifth century BC.

87 What is the precession of equinoxes?

The Earth's axis of rotation is not perpendicular to its orbital plane, but tilted by about 23°. Over a short period, it can be considered fixed, pointing to the same direction in space – the North Pole now points almost exactly at the star *Polaris*. However, the axis is actually slowly changing its alignment while keeping the same tilt with respect to the Earth's orbit, like a top or gyroscope that wobbles as it spins. This wobbling motion of Earth is called *precession*.

If the Earth were perfectly spherical, its axis of rotation would be stationary – no precession would take place. It is because our planet bulges at the equator that it behaves like a gyroscope. The attraction of the Sun on the swollen equator tends to pull it back into the orbital plane, and this makes the axis of Earth's rotation shift in

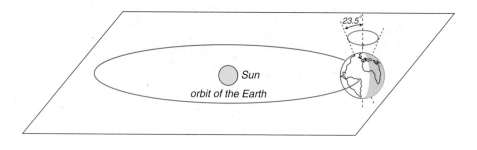

the direction perpendicular to this axis and to the force. The same phenomenon occurs under the influence of the Moon and, to a lesser degree, of the other planets. This shifting motion of Earth's axis of rotation traces a cone in space over time, completing one full cycle in 26 000 years.

The shifting motion is more properly called the "precession of the equinoxes" because it has the effect of displacing the succession of the seasons along the Earth's orbit. This simply means that, as seen from Earth, the position of the Sun at a given moment in a season changes with respect to the background stars. For example, 2000 years ago at the spring equinox, the Sun was rising in the constellation *Aries*. On the equinox today, it rises in the constellation *Pisces*.

Another consequence of precession is that the celestial pole (i.e. the direction in which the Earth's axis points) drifts with time. Nowadays it points to the bright star *Polaris*, whereas 4000 years ago, at the time of the Mesopotamian astronomers, it pointed at a different star, *Alpha Draco*, in the constellation *Draco*), and 5 500 years from now our new "Polaris" will be *Alpha Cephei*.

88 What caused the "ice ages" on Earth?

Glaciers currently cover 10% of the emerged lands on Earth but 18 000 years ago, at the peak of the last ice age, they covered 32%. The British Isles, northern Europe, Canada, and the northern regions of the United States were all buried under thick layers of ice.

If we go back even further in time, we see that on a timescale of tens of millions of years Earth's climate oscillates between periods of cold and warm, the latter sometimes associated with the complete disappearance of ice fields. On this timescale, we are in a relatively cold period that began about 60 million years ago. Within this cold period, however, temperature has fluctuated significantly over periods of 10 000 to 100 000 years. The colder periods, when polar caps were very extensive, are known as "glaciations" or "ice ages." The warmer periods alternating with the glaciations are

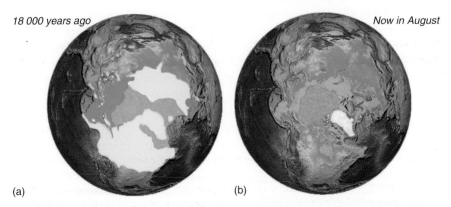

18 000 years ago   Now in August

(a)   (b)

(a) In white, the extent of glaciation in the northern hemisphere 18 000 years ago.
(b) Present conditions during summer. Ice floes are shown in blue gray, emerged land in green. Credit: M. McCaffrey NGDC/NOAA.

called "interglacials." We are in an interglacial period now. The last ice age peaked about 18 000 years ago and ended 10 000 years ago.

How can we know all this? The last glaciation left highly visible traces on the land: U-shaped glacial valleys, moraines of glacial debris, and a great many lakes, some of them enormous (the Great Lakes of North America, the Aral Sea). At the same time, it erased all vestiges of previous glaciations. However, it is still possible to date previous ice ages by analyzing seabed material. Water molecules containing the O^{16} isotope,[†] being lighter than O^{18}, evaporate faster. During non-glacial periods, water rich in O^{16} that evaporated from the oceans returns to the sea as rain, while, during ice ages, some of the O^{16} water falls as snow at the poles and is locked into polar ice fields, leaving the seas enriched in O^{18}. When small, shelly planktonic animals (foraminifera) die, sink to the bottom of the sea, and accumulate as sediments, their shells retain the relative content of O^{18} to O^{16} present in the seawater when they were alive. By analyzing sediment cores obtained through deep sea drilling, we can thus reconstruct past variations in temperature.

How do we explain the observed variations? Not very well – complex phenomena and interactions are involved, and we cannot yet answer this question with full certainty. Four factors are of particular importance:

- continental drift, which changes the positions of continents and oceans with respect to the poles and the equator, thus affecting the total amount of heat absorbed from the Sun;
- the uprising of large mountain ranges (Himalayas, Andes) and high plateaus (Tibet, North America), which modifies both the circulation and temperature of the atmosphere;
- the carbon dioxide content of the atmosphere, which produces the greenhouse effect (Q. 92) and which varies as a function of volcanic activity and the number and type of living organisms in existence;
- finally, astronomical causes that modify the amount of sunlight received by the Earth and the relative intensity of the seasons.

Of all these factors, the first three are the most important and could explain our long-term climatic changes with no help from the fourth. On the other hand, astronomical events do correlate relatively well with the glaciations that have taken place over the last 400 000 years.[‡]

[†] The *isotope* of a chemical element is an atom whose nucleus has the same number of protons; that is, the same atomic number, but a different number of neutrons, hence a different atomic mass. Hydrogen, for example, has three isotopes: regular hydrogen, deuterium, and tritium. All three atoms have only one proton each in their nuclei, but deuterium also contains one neutron, and tritium has two, whereas the far more common hydrogen nucleus contains no neutrons. Carbon-14 is an isotope of the more common carbon atom, carbon-12. Both atoms have 6 protons in their nucleus, but C^{14} contains 8 neutrons, while C^{12} has only 6 of them.

[‡] The astronomical explanation was proposed in 1842 by the French mathematician Joseph Adhémar, then studied in detail by the Serbian astronomer Milutin Milankovitch in the first half of the twentieth century.

Variations in the tilt of the Earth's axis influence our overall exposure to the Sun and the intensity of our seasons. This tilt oscillates between 21.5° and 24.5° over a period of 41 000 years (it is 23.4° at present). The precession of the equinoxes (Q. 87), with a period of 26 000 years, changes the month of the year in which the Earth is the furthest away from the Sun (winters are colder when they peak just when the Earth is most distant from the Sun). Finally, the eccentricity of the Earth's orbit changes from 0 to 6% over a period of 100 000 years, modifying our annual exposure to the Sun (the Earth receives less energy per year when its orbit is more circular).

The periodicities of the recent glaciations show all these astronomical cycles in play, suggesting that they either cause them or at least serve as triggers to set them off. Other mechanisms, such as periods of intense volcanism or large-scale interruptions in the main ocean currents, probably help to amplify the purely astronomical effects.

Since, in the short term, the variations in Earth's orbit correlate well with glaciations, can we use them to predict the next ice age? Some authors say that we are already overdue for one, while others predict that the current inter-glacial period will last for another 50 000 years [5]. The answer is difficult to come by. On the one hand, we do not yet understand the mechanism by which astronomical events affect climate, and on the other hand, effects related to human activities (Q. 92) have also become important [26].

89 What causes the Earth's magnetic field?

Why the Earth has one is still something of a mystery. Its field is similar to one that a bar magnet placed inside our planet would produce, with the classic lines of force arcing between the north and south magnetic poles. Although there is a great deal of iron in the Earth's interior, however, it cannot be permanently magnetized because the very high temperatures prevailing there would make any magnetized substance lose its magnetism.[†]

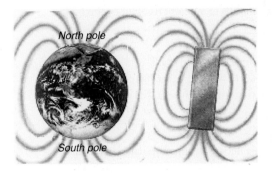

The Earth's magnetic field resembles that of a bar magnet.

The only other way to produce such a field is with moving electrical currents. Scientists now think that the Earth's field may be the result of convection currents in the viscous metal masses inside the planet (Q. 82). These would, in effect, be acting as a dynamo, creating the weak field we detect at the surface.

[†] The temperature at which a substance loses its magnetism is called the *Curie point*. The Curie point of most substances is below 550 °C, a temperature that is reached at about 30 km beneath the Earth's surface.

90 Does the Earth's magnetism affect people?

According to certain popular beliefs and the principles of Chinese Feng Shui, you will sleep better if your head is to the north.

The idea apparently originated with ancient Chinese practitioners who noticed that a thin iron needle floating on water would point north (hence their invention of the compass). While it is true that some animals make use of the Earth's weak magnetic field to find their ways during migrations, they can do so because their brains contain magnetite, an iron oxide. Lacking a similar concentration of magnetically sensitive substances in our own brains, we cannot react to the Earth's weak field – not even in our sleep.

That does not mean that the magnetic field around Earth is of no importance to us, however. Without it, we might not even exist. It is the magnetic field enveloping Earth that protects us from the deadly showers of high-energy particles ejected by the active Sun (Q. 118).

91 Why is the magnetic north different from the geographic north?

The magnetic needle of a compass almost never points towards the geographic, or true, north.[†] The reason for this discrepancy is that Earth's magnetic poles do not coincide with its geographical poles (the points through which our planet's axis of rotation passes). The two points are actually quite far apart.

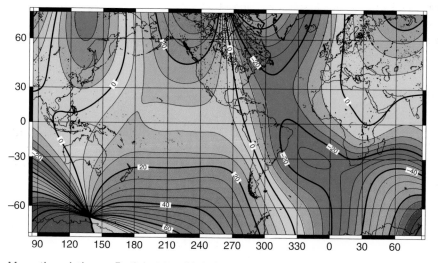

Magnetic variation on Earth in 2005. Variation can be as great as 30° west (red) or east (blue). Credit: USGS.

[†] The angular difference between the directions of the magnetic and geographical poles is called the magnetic *declination* (by physicists) or *variation* (by sailors and aviators).

The north magnetic pole is currently located in the Canadian Arctic at a latitude of +81°, about 800 km from the geographic North Pole, and it is moving ever further north at about 40 km/yr. The south magnetic pole is located in the Antarctic Ocean at a latitude of −64°, about 2400 km from the south geographical pole.

The Earth's magnetic axis is not aligned with its rotation axis because the movements of the semi-liquid iron in its interior (Q. 89) are not perfectly coaxial with the planet's axis of rotation. The liquid metal's axis varies over time and the magnetic polarity can even reverse itself. That event occurs about once every 300 000 years on average, but the last time it happened was 740 000 years ago, making some geologists think that we are more than a little overdue ...

When magnetic polarity does reverse itself it happens rather quickly, over a period of just a few thousand years. While this is going on, the Earth temporarily loses the protective magnetic shield that normally deflects the Sun's lethal particles (Q. 118), with potentially tragic consequences for the animals and plants that inhabit our world.

92 What is the greenhouse effect?

A greenhouse is warm because the sunlight shining through the glass panes during the day heats the interior, and most of this heat remains trapped in the air, the plants, and the ground there. Two phenomena are responsible for this: (a) the warm air cannot escape by convection because it is blocked by the glass panes, and (b) glass has the property of letting sunlight (visible light) penetrate it to come in, but not letting the heat radiation (which is infrared radiation) pass through to get out. In other words, it is an efficient heat trap. A car left in the sun with its windows closed collects heat in exactly the same way: the temperature inside becomes much hotter than the air outside.

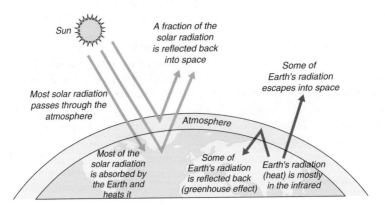

The carbon dioxide molecules tend to reflect back the heat radiated from the ground, making the atmosphere work like a blanket that keeps the Earth from cooling excessively.

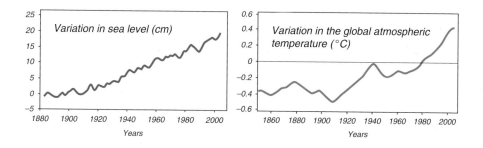

Of the two phenomena, the most effective in a greenhouse is actually the suppression of convection, but the expression "greenhouse effect" is now often used to describe the second mechanism, the trapping of infrared radiation.

This trapping effect works on Earth because the atmosphere acts similarly to the glass panes of a greenhouse. Our atmosphere is made up mainly of nitrogen (78%) and oxygen (21%), with traces of other gases. The atoms in the molecules of oxygen and nitrogen (O_2 and N_2), are strongly bound and do not react when struck by photons of infrared radiation. Unlike the oxygen and nitrogen, however, several other gases in the atmosphere, even in very small quantities, can absorb and re-emit significant amounts of such infrared radiation (heat). Carbon dioxide (CO_2), essentially, but also water vapor (H_2O), methane (CH_4), and several nitrogen oxides (NO_x), these constitute the greenhouse gases. Their molecules, composed of three weakly bound atoms, easily absorb infrared radiation, begin vibrating, then re-emit it. This re-emitted heat goes off in all directions, part of it escaping into space but part of it returning to Earth. Although the result is similar to the effect produced by a glass pane, the mechanism is different: a glass pane absorbs the heat (and becomes warm), while the greenhouse gases re-emit a certain fraction of the heat they receive.

The greenhouse effect is actually a good thing for us. Without it, much of the energy we receive from the Sun would escape into outer space and the average temperature at the surface of our planet would drop to about $-20\,°C$. Instead, thanks to this trapping effect, the average temperature is $15\,°C$, which prevents the oceans from freezing and makes Earth hospitable to life.

The problem is that the benefit we receive from the greenhouse effect is the result of a delicate balance. When there is too much carbon dioxide, overheating occurs. This is what happened to our sister planet, Venus, whose atmosphere is almost entirely made up of carbon dioxide. Conditions there are hellish, with a surface temperature of $470\,°C$. For this reason, the increasing amounts of carbon dioxide in our own atmosphere represent a real danger. The carbon dioxide content has augmented by more than 30% in 150 years due to human activities, and the impact is already noticeable. In the last century and a half, the average sea level has risen 20 cm due to the melting of glaciers [17], and the average temperature[†] at the surface of the Earth has increased by $1\,°C$ [9].

[†] This is the temperature "anomaly," the difference between observed temperatures and the long-term mean.

93 Have days on Earth always been the same length?

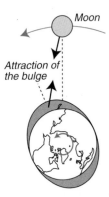

Moon

Attraction of the bulge

Days on Earth are getting longer and longer, and the Moon is to blame. The attraction of the Moon creates tides in our seas, but also acts on the Earth's solid crust, albeit to a lesser degree. The tidal rise of the terrestrial crust amounts to about 20 cm on average. Combined with the centrifugal force of the tandem Earth–Moon system, the action of the Moon on the Earth manifests itself by a swelling on the side facing the Moon and also on the opposite side of Earth (Q. 107). This bulge follows the Moon as it revolves around the Earth (1 revolution in 27 days), but meanwhile the Earth is rotating, too, (one turn in 23 h 56 min, see Q. 94), carrying the tidal bulge with it. Thus the bulge is always slightly ahead of the Moon, tugs it forward, tending to accelerate it. Conversely, this slight deflection of the attraction of the Moon slows the Earth's rotation. The day is currently lengthening by about 1.8 ms/century.

Computer simulations have shown that when the Earth was newborn, 4.5 billion years ago, a day was only 6 h long. Paleontological measurements of daily growth in 400 million years old fossilized corals show that, by then, the day had lengthened to 22 h. In theory, several billion years from now, the Earth and the Moon would be locked face to face and their rotations would be synchronous. An Earth day would then be as long as one lunar month, about 50 of our present days. This will never actually come to pass, however, because long before that the Sun will become a red giant and probably destroy both the Earth and its satellite (Q. 39).

94 What is sidereal time?

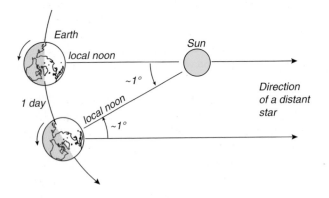

In day-to-day life, the Sun defines time for us. Noontime, locally, happens when the Sun crosses the meridian.[†] And when it crosses that meridian once again, one day has passed. However, since the Earth completes a revolution around the Sun (360°) in one

---

[†] The meridian of a given location corresponds to an imaginary great circle, traced on the surface of the Earth or on the sky, that crosses the local vertical plane and the polar axis. The meridians on Earth are defined with respect to the reference meridian that goes through the Greenwich Observatory, near London, England (Q. 98).

year (about 365 days), during that same day it has moved along its orbit by an angle equal to 360°/365, or approximately 1°.

Now if, instead of the Sun, a distant star is observed crossing the meridian at midnight, it would not cross the meridian at midnight the next day, but a little earlier, with the time difference being equal to the time it takes the Earth to cover about 1° of its 360° orbit, i.e. 24 h/360, or almost 4 min – more precisely, 3 min and 56 s. This means that the sidereal day, defined as the time between successive crossings of the meridian by a distant star, lasts 23 h, 56 min, 4 s, which is the true rotation period of the Earth. And it is *sidereal time*, the "clock of the stars," that astronomers (except for solar astronomers) use to point and track stars in the sky.

95 Why is the day divided into 24 hours?

Our numeral system today is a decimal one (based on 10), its origin most likely linked to the use of the fingers of both hands for counting purposes. The Egyptians, who created the first mathematical division of the day, used a duodecimal system (based on 12). That may be because there are approximately 12 lunar cycles in a year or, as some have suggested, because they counted using the joints of their fingers – excluding the thumb, which served as the "pointer."

To tally the passage of time during the day, Egyptian sundials were marked with 12 divisions of equal length between sunrise and sunset. At night, astronomers would time their observations by the rise of a given star or with a clepsydra (a water clock) and would also divide nighttime into 12 equal periods. This is the origin of our 24 h day.[†]

The one obvious problem with this system is that, although days and nights are the same length at the equinoxes, the 12th part of a long summer day is larger than the 12th part of a short winter day. That meant that the length of an hour had to vary with the seasons. And, in fact, it was not until the fourteenth century, when the first mechanical clocks were constructed, that the hours were made equal for daytime and nighttime and constant throughout the year [16].

As for minutes and seconds, we can thank Ptolemy, the Babylonians, and the Sumerians. When working on his *Geography*, Ptolemy felt the need to divide the angular degree into smaller units in order to describe the locations of geographical sites more precisely in terms of latitude ("breadth" in Latin) and longitude ("length" in Latin). He divided the degree into 60 divisions, calling that unit angle the "small part" (*partes minutae* in Latin). For even finer measurements, he divided this small angle again into 60, and to distinguish the two small units, he called the first one the "first small part" (*partes minutae primae* in Latin), and the second one "second small part" (*partes minutae secondae* in Latin). Translators later managed to truncate the two expressions: *minute* was kept for the first small angle and *second* for the smaller unit.

Why a division into 60? This came from the sexagesimal system that the Babylonians used in their astronomical observations, a system that they themselves had inherited

---

[†] The word "hour" comes from the Egyptian *har*, meaning "day" or "trace of the Sun in the sky," and the Egyptian god of the dawn was named Horus.

from the Sumerians, who invented it around 2000 BC. The reason why the Sumerians were moved to develop a sexagesimal system remains obscure, but it has certain advantages: 60 is useful when dealing with fractions, being divisible by the numbers 1 through 6, and also by 10, 12, 15, 20, and 30; it is also a submultiple of the approximate number of days in a year (360). The division of a circle into 360 degrees undoubtedly originated in the number of days per year, with one degree corresponding roughly to the daily displacement of the Sun against the star background.

Subsequently, because angles in the sky are related to time via the Earth's 24 h rotation period (1 h = 360/24 = 15°), it was natural to divide time in the same way, bringing us the minute and second of time.

96 How do sundials work?

The principle behind the sundial is very simple; as the Earth rotates, the Sun appears to move across the sky, and the shadow projected by any object also moves. From this we can determine the time of day. The object used to project the shadow, an upright post or obelisk for the Egyptians, is called a *gnomon*, Greek for "indicator" or "that which reveals."

Position of an obelisk's shadow at different times of day.

The problem with an elementary sundial of this type is that, since the Earth's axis is inclined, the path of the Sun in the sky – hence, the shadow tracing that path –varies with the seasons. The problem can be eliminated by tilting the gnomon to make it parallel to the Earth's axis of rotation, i.e. by pointing it at the celestial pole. The tilted gnomon is called a *stylus*. The hour markings on such a sundial are valid all year round.

Nevertheless, with a horizontal sundial (on the ground, for example) or a vertical one (on a wall), the shadow projected by the gnomon does not move with uniform speed and the angles separating the hourly marks are not constant, the intervals increasing when the Sun is closer to the horizon. To eliminate this last difficulty, the plane of the sundial should be tilted so as to make it parallel to the equator, i.e. perpendicular to the gnomon. The angles between the hours are then all equal to 360°/24 = 15°. Two gnomons

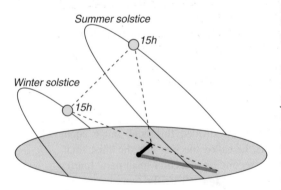

With a gnomon parallel to the Earth's axis of rotation, the projected shadow (here shown marking 3 p.m.) varies in length but is always at the same position, regardless of the time of year.

are needed for this system, however; one above for fall and winter and one below for spring and summer (in the northern hemisphere). This type of sundial is called "equatorial."

In all cases, the *length* of the projected shadow on a sundial (not its *position*) depends on the season, since the Sun is higher in the sky in summer and lower in winter. This makes it possible to use a sundial to derive the calendar date, at least approximately. For this purpose, curved lines are sometimes carved into sundials: the tip of the gnomon shadow's with respect to the curves indicates the approximate day in the year – or at least the month.

Equatorial sundial in Beijing.

A sundial shows the *local* time, and that depends upon the longitude of the site: the shadow indicates noon when the Sun crosses the meridian. In practice, the time we use in everyday life is *civil* time, which is defined by the time zone for any given location (Q. 98), adjusted if needed for daylight or standard time.

97 How can the Sun be used to find directions?

Since a correctly oriented sundial can show the time, then, conversely, knowing the time of day and the direction of the Sun in the sky, one can find one's bearings.

First of all, when it is noon (in the northern hemisphere), we know that the Sun is to the south in the sky. Note that "noon" in this case means solar noon; there may be a difference between civil time and local solar time, and that must be taken into account. If it is not solar noon, any traditional (non-digital) watchface can be used as a sundial – hence as a compass. The principle is based on the fact that the small hour hand of a watch rotates twice as fast as the Sun, since it goes twice around the clock face during a 24 h day.

Here is how you proceed. First, find solar time, mentally adjusting the time shown on your watch face if necessary, e.g. in most of the United States, subtracting one hour for summer daylight saving time. Then, point the (mentally adjusted) hour hand towards the Sun. Since that hand advances twice as fast as the Sun, south will be in the direction of the bisector between the hand and the number 12 on the clock face. In the southern hemisphere, it is the numeral 12 on the clock face that is pointed at the Sun; north is in the direction of the bisector between the small hand and the number 12.

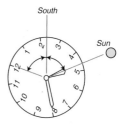

This method is only approximate, as it makes no correction for the longitude of the site. It is also difficult to use when the Sun is high in the sky, especially if you are near the equator.

98 How was the time zone system established?

In the past, the time of day was locally defined, usually by means of a sundial. Consequently, it varied from town to town as one traveled in the east–west direction. But the slight, incremental differences in the neighboring areas seldom mattered because travel was a slow affair. Then came the railroads, and those time discrepancies became a nightmare.

By 1840, the British railroads had decided to adopt one single, official time for their entire network. That was obviously impossible for the US and Canada, however, with territories covering 55° of longitude and 3.5 h of time difference. How could travelers know how long a trip was going to last if the time shifted with every mile traveled? How could a train schedule be written? How could train connections be coordinated for different legs of a journey involving different railroad companies unless both companies ran on the same time system? So in 1883, at the instigation of the American railroad companies, the US decided to officially divide the country into four time zones. And the following year, an international conference meeting in Washington, DC, extended the system to divide the entire world into 24 zones, each one 15° of longitude in width.

The meridian passing through Greenwich, England, was adopted as the reference point, with the mean solar time of Greenwich (Greenwich Mean Time, GMT) furnishing "time zero."[†]

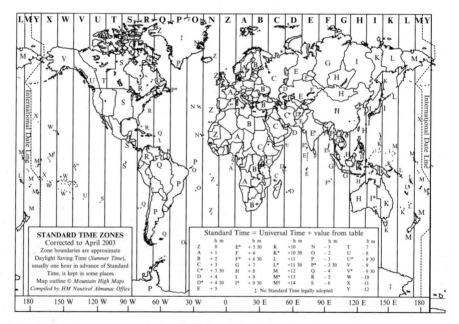

STANDARD TIME ZONES
Corrected to April 2003
Zone boundaries are approximate
Daylight Saving Time (*Summer Time*),
usually one hour in advance of Standard
Time, is kept in some places
Map outline © *Mountain High Maps*
Compiled by HM Nautical Almanac Office

Standard Time = Universal Time + value from table

	h m		h m		h m		h m		h m
Z	0	E*	+ 5 30	K	+10	N	− 1	T	− 7
A	+ 1	F	+ 6	K*	+10 30	O	− 2	U	− 8
B	+ 2	F*	+ 6 30	L	+11	P	− 3	U*	− 8 30
C	+ 3	G	+ 7	L*	+11 30	P*	− 3 30	V	− 9
C*	+ 3 30	H	+ 8	M	+12	Q	− 4	V*	− 9 30
D	+ 4	I	+ 9	M*	+13	R	− 5	W	−10
D*	+ 4 30	I*	+ 9 30	M†	+14	S	− 6	X	−11
E	+ 5			‡ No Standard Time legally adopted				Y	−12

The 24 time zones and the choices made by each country. Note that certain countries have adopted a half-hour step. Credit: HM Nautical Almanac Office.

[†] This decision was a mighty disappointment to the French, who had hoped to have the meridian of Paris adopted. However, the choice was almost inevitable since, at that time, 75% of all existing nautical charts were already based on the Greenwich meridian. So the French gave in but asked for a promise in exchange: that the English-speaking world would adopt the metric system. A hundred years have gone by and, with the US still holding out, the promise has only been partially kept.

The Moon

99 How did the Moon form?

Until man landed on the Moon, there were three main theories for its origin. The first, *simultaneous formation*, proposed that it formed from the same dust-gas protodisk as the Earth. The second, *fission*, proposed that the very young Earth was liquid and that centrifugal force had caused a portion of it to become detached, forming our satellite. According to the third theory, *capture*, the Moon formed independently in the solar system and then was captured by Earth's gravity.

Analysis of the lunar rock samples brought back by the *Apollo* astronauts in the late 1960s and 1970s invalidated these theories. The chemical composition of lunar rocks is different from those of Earth, in particular their iron content. Therefore, the Moon could not have been pinched off from a liquid Earth or been formed from the same portion of the solar nebula. Neither could it have been formed elsewhere in the solar system and then been captured, since, despite significant differences in composition, lunar material shows striking similarities with that of the Earth.

Currently, the most probable scenario is that of a collision. It would have occurred about 4.6 billion years ago when the young planets had just completed their *differentiation* (Q. 52); that is, their metal components (especially iron) had just coalesced into a central core surrounded by a rocky mantle. Another small protoplanet the size of Mars then collided with Earth and a large amount of material from the external layers (mantles) of both bodies was ejected into space. This debris was flung into orbit around Earth and later aggregated to form the Moon. The ferrous nucleus of the colliding body was shock-absorbed into the Earth and swallowed up into its core, which explains why the Moon contains so little iron.

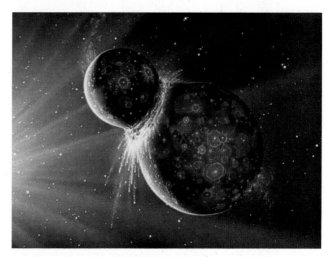

The Moon probably formed as the result of a collision between the young Earth and an object the size of Mars. Credit: NASA.

This theory has been confirmed by computer simulations which successfully describe the differences and similarities between the Earth and the Moon. Moreover, since collisions between two such relatively large bodies must be rare, this theory explains the oddity of the Moon's large size compared to other satellites in the solar system. For all other planets, the ratio of the mass of the satellites to that of their planets is at most 0.025%, whereas the mass of the Moon represents 1.23% that of Earth. The exceptional size of the Moon suggests that it did not form like the other satellites, while it is well explained by a rare event such as a gigantic collision.

100 Why is the Moon covered with craters?

The Moon's surface is riddled with craters of all sizes. This view shows a portion of the crater Ptolemy (outlined for clarity), which is 160 km in diameter, and crater Herschel, 40 km in diameter. The photo was taken by the *Apollo 12* astronaut who stayed in orbit after the separation of the lunar module (visible at top on its descent towards the Moon's surface). Credit: NASA.

Unlike craters on Earth, almost all lunar craters are non-volcanic in origin, the result of meteorite impacts. They range in size from a few millimeters to more than 200 km in diameter, with some few, called "basins," attaining diameters of up to 2000 km.

Most lunar craters were created at a time of intense meteorite traffic in the early solar system (Q. 33), about 4 billion years ago. The Earth was not spared at that time and was bombarded as well. Why are there no terrestrial impact craters like those on the Moon? On Earth such craters, except for the youngest ones, have been erased by water and wind erosion and by geological activity – volcanoes and tectonic movements renew the surface continuously. As a result, very few terrestrial impact craters remain visible.

The Moon is geologically inert now; nothing erodes its surface but slow, continuous meteoritic bombardment. It looks much the same today as it did 3 billion years ago.

Meteor Crater in Arizona, 1 km in diameter, was formed by the impact of a 30 m meteorite about 50 000 years ago. This young terrestrial impact crater resembles Moon craters that are much older. Credit: US Army Air Service.

101 What are the large dark areas on the Moon?

The Moon photographed by the *Apollo 11* astronauts on their return flight to Earth in July 1969. They had landed in *Mare Tranquillitatis*. The dark zones, the "seas," are regions flooded by massive ancient lava flows. Credit: NASA.

The flat dark regions of the Moon were called "seas" (or *maria*, sing. *mare*) by seventeenth-century astronomers who thought they were the beds of ancient oceans. We now know that they are the remains of large-scale basaltic lava flows that filled the floors of immense impact craters. In some cases the lava was confined inside the crater, and the sea appears circular (e.g. *Mare Crisium*). In other instances the lava flow spilled out through gaps in the sides of craters to cover adjacent craters (e.g. *Mare Serenitatis*). Lunar seas look dark because of the blackish color of basalt.

The lava originated from hot magma rising through fractures created in the crater floors at the time of impact. Radioactive decay heating the mantle triggered the flows about 3.7 billion years ago, that is, a few hundred million years after the end of the era of intense bombardment. This would explain why the seas are not covered with new craters.

102 What does the far side of the Moon look like?

The Moon, being in synchronous rotation with the Earth (Q. 105), always shows us the same side – until recently its "far side" was a complete mystery. Some imaginative souls

The far side of the Moon as seen by the astronauts of *Apollo 16* in 1972. It is pockmarked with so many more craters than the near side that it has been called the "disfigured side." Credit: NASA.

even used to claim that it was inhabited by an ancient civilization or that aliens had a base there, hidden from the prying eyes of Earthlings.

But with the coming of the space age, the Moon, which is the second most studied of all bodies in the solar system (after Earth), has revealed most of its secrets. The veil of mystery over the far side began to be lifted in 1959, when it was photographed by the Soviet probe, *Luna 3*. Since then, several other probes have photographed it, and humans finally saw it with their own eyes when the *Apollo 8* astronauts orbited the Moon in December 1968.

Surprisingly, the far side looks very different from the side facing Earth: it is much more heavily cratered; or rather, it has fewer *maria* (Q.101), features which cover only 2.5% of its surface versus 31% on the near side. This may be the result of a slight asymmetry in the Earth's gravitational influence: the crust is thicker on the far side, making it harder for molten magma from the interior to rise to the surface and form the smooth maria.

103 Does the Moon have the same composition as the Earth?

If we could take a scenic walk on the Moon, we would be struck by how dark its surface is. The full Moon appears bright, but the lunar surface has the color and reflective power of asphalt.[†] Most of the rocks on Earth are sedimentary, formed from light-colored shells and coral weathered into particles by water and wind erosion, then recemented by pressure. On the Moon, most rocks are volcanic. The floors of lunar maria are composed of basalt from lava flows (Q. 101), which are dark, like the rocks of the

---

[†] Overall, the Moon reflects only about 10% of the sunlight striking it (a ratio called the *albedo*), while the Earth reflects about 37%. This is due to the lighter color of the rocks, but also to the presence of clouds, seas, snow, ice, and vegetation.

A large lunar boulder photographed during the *Apollo 17* mission in 1972. This boulder may have rolled down from a hill close by. A layer of regolith covers the surrounding land. Credit: NASA/JSC.

Hawaiian archipelago, while the elevated areas are mainly *anorthosite*, a volcanic rock similar to basalt, less dense and lighter in color, but still darker than most Earth rocks.

There are few outcrops of rock on the surface of the Moon, only occasional large boulders. Instead, the surface is covered by a thick layer of dust called *regolith*, created by the pulverization of base rock by meteoritic bombardment. The thickness of the regolith varies between 2 and 20 m. A walker's foot sinks into it as into moderately soft snow. Prior to the first manned Moon mission (*Apollo 11*), it was feared that the lunar module would sink deeply into this fine powder and that astronauts would have difficulty moving around, but that proved not to be the case.

What is the interior of the Moon like? Just as for Earth (Q. 82), the Moon's interior can be divided into three zones: the crust, the mantle, and the core. The lunar crust is 70 km thick and consists mainly of anorthosite. Below that is a 1500 km-thick mantle composed of minerals such as olivine and pyroxenes. At the center is an iron and nickel core that is much smaller than Earth's, with a diameter of 300 km and a mass of only 4% of that of the entire Moon, as opposed to 30% in the case of Earth.

Neil Armstrong's bootprint. Made during man's first walk on the Moon in 1969, it should last for a million years. There is no wind or rain to erase it.

104 Why does the Moon lack an atmosphere?

The Moon cannot maintain an atmosphere simply because its gravity is not strong enough to prevent the escape of gases. Some planets also have had this problem, whereas others have retained large atmospheres. It all depends on a body's surface temperature and strength of gravity (Q. 51).

The Moon's gravity is only one sixth that of Earth, its radius only a quarter. It is thus much easier to escape the Moon's gravitational pull than the Earth's (the lunar escape velocity is 2.4 km/s, compared to 11.2 km/s for Earth). With lunar daytime temperatures reaching 120 °C, gas molecules bounce around at high speed and lighter gases such as hydrogen and helium escape into space. Since there are no significant sources of heavier gases, no real atmosphere can develop and persist on the Moon. There are traces of hydrogen, helium, argon, and other gases coming from the solar wind or caused by the meteoritic shocks heating lunar rocks, but the quantities are minute. Atmospheric pressure at the surface is only 10^{-15} that of Earth.

105 Why does the Moon always present the same face to Earth?

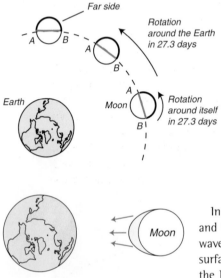

The Moon always shows us the same face because it spins once around on its axis in *exactly* the same amount of time as it takes to revolve around Earth: 27.3 days.[†] This is no coincidence; it is a tidal effect. But in this case, the tide is created by the Earth on the Moon.

The Earth, being 80 times more massive than the Moon, creates much stronger tidal forces on the Moon than the Moon causes on Earth. There is no water on the Moon, but these tidal forces are strong enough to change the Moon's shape, making it bulge out on the side facing Earth, an effect called a *solid tide*.

In the past, the Moon used to spin faster on its axis and the bulge moved across its surface just as the tide wave does on Earth. This continual distortion of the surface led to a loss of energy through friction, causing the Moon to slow down progressively. When the Moon was spinning once on its axis in precisely the time it took to revolve around the Earth, the process stopped and the Moon's bulge became stationary, with the bulge facing us.

This situation is not unique to our Moon. Most satellites of other planets (notably those of Mars, Jupiter, and Saturn) also always show the same face to their planets. Only satellites that are far enough away from their planet to be essentially free of tidal effects escape this fate.

What happened to the Moon is happening to us. Our daily tides are causing the Earth to lose energy through friction between the ocean waves and seashores, the Earth is

---

[†] The period between two full Moons is a little longer, 29.5 days, because Earth has also revolved around the Sun during that time.

rotating more and more slowly, and the days are getting longer by about 1.8 millisecond per century (Q. 93).

106 Why does the Moon, rather than the Sun, cause most of our tides?

The Moon is the main cause of our tides. This is true of the coasts of Europe, the Atlantic and Pacific coasts of North America and, indeed, of most of the coasts on Earth.

Yet the Sun's attraction is certainly stronger. We revolve around the Sun, not around the Moon. Although 390 times more distant than the Moon, the Sun is 27 million times more massive. According to Newton's law, the attraction between two bodies is proportional to the product of their masses divided by the square of their distance (Q. 80). The attraction of the Sun is therefore $27 \cdot 10^6/390^2 = 178$ times stronger than that of the Moon.

Still, we do not fall into the Sun. The reason is that its attraction is compensated by the centrifugal force acting on the Earth as it revolves around it.

So what counts is not the attraction itself but the *difference* of attraction as a function of the distance to the Sun (or the Moon). For example, a point on the Earth's surface facing the Sun will be slightly more attracted than a point on the opposite side of the Earth. Instead of varying as the *square* of the distance (as attraction does), the differential attraction varies as the *cube* of the distance,[†] so that the tide caused by the Sun is about half of that due to the Moon ($27 \cdot 10^6/390^3 = 0.45$). Overall, then, two thirds of the tide is due to the Moon and one third to the Sun.

Although this is true in general, the presence of continents and local coastal configurations can diminish or amplify tidal effects, so that in certain places the tidal component due to the Moon will overwhelm that of the Sun, or, conversely, become negligible.

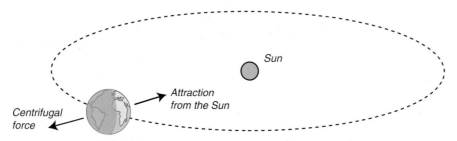

The attraction of the Sun is compensated by the centrifugal force acting on the Earth during its annual revolution. The solar tide is not due to the Sun's attraction per se, but to the differences in the strength of its attraction at various points on Earth. The same is true for the lunar tide.

[†] The derivative of $1/d^2$ is $-2/d^3$.

107 If the tide is mainly caused by the attraction of the Moon, why is there a simultaneous high tide on the Earth's opposite side, facing away from the Moon?

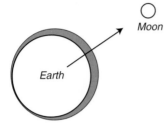

Since the tide is primarily due to the attraction of the Moon, one would expect that the waters of the oceans would be drawn to the side of Earth that faces the Moon, as in the figure on the right. Then there would only be one tide a day, not the usual two.

What actually happens is that the Moon does not turn exactly around the center of the Earth, but around the center of gravity of the Earth–Moon couple. The same is also true of the Earth going around the Sun. It is the center of gravity of the two bodies combined, Earth + Moon, that revolves around the Sun, while the Earth describes little undulations on either side of the perfect orbit.

Since the Earth is much heavier than the Moon, the center of gravity of the couple Earth–Moon is located near the Earth's center, about 1700 km below the surface.[†] In one lunar month, the Earth thus revolves around this point and, like all revolving bodies, it is subjected to centrifugal force.

At the center of Earth, the centrifugal force exactly counteracts the attraction of the Moon on the entire Earth (the distance between the Moon and the Earth has adjusted

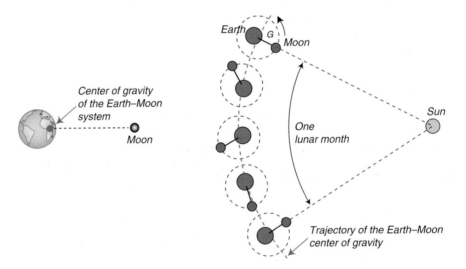

In one lunar month, the Moon does not turn around the center of the Earth, but around the center of gravity of the Earth–Moon system, and the Earth does the same.

[†] The mass of the Earth is 81 times that of the Moon, so the center of gravity of the couple Earth–Moon is at 1/80th of the distance between Earth and the Moon. Since the distance from Earth to the Moon is 384 000 km, the center of gravity of the Earth–Moon couple is 4700 km (384 000/81) from the center of the Earth, or about 1700 km below the surface of the Earth.

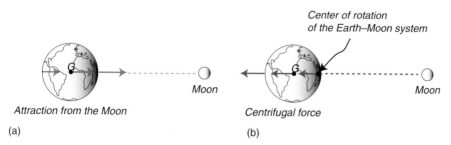

Two forces combine to create the tides: (a) the gravitational attraction of the Moon, and (b) the centrifugal force due to the Earth's rotation around the Earth + Moon center of gravity.

itself so as to create this equilibrium). But this is not true for other points on Earth, for the attraction of the Moon varies with the distance to it. This attraction is stronger on the side of Earth where the Moon happens to be, and less strong on Earth's opposite side.

So, ultimately, lunar attraction overcomes the centrifugal force on the side of Earth facing the Moon and this creates the predictable bulge of water. On Earth's other side, it is the centrifugal force that overcomes the lunar attraction and creates the second tide [10, 33].

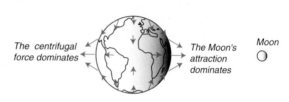

108 Is it just coincidence that the apparent diameters of the Moon and the Sun are the same?

The diameter of the Sun is about 400 times that of the Moon, but the Sun is also 400 times further away from us than the Moon. The configuration leads to the same apparent diameters for both objects when seen from Earth (0.5°).

This coincidence grants us the opportunity to have total eclipses of our star, a majestic cosmic phenomenon that used to terrify ancient peoples.

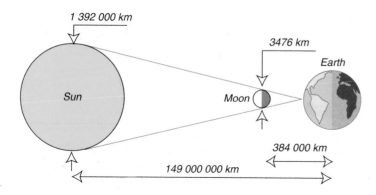

But this matching of apparent diameters did not always exist. A few million years ago, the Moon was closer to Earth and covered an apparent angle larger than the Sun. And millions of years from now, the Moon will have receded, and total eclipses will no longer be possible.

109 How often do solar eclipses occur?

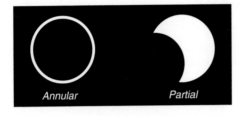

Annular   Partial

Solar eclipses are relatively frequent. There are on average 2.4 eclipses of all types (total, annular, and partial) per year, with a minimum of 2 and a maximum of 5 [18].

When the apparent disk of the Moon completely hides that of the Sun, the eclipse is referred to as *total*. Only the more extended chromosphere and corona remain visible, showing spectacular hues of silver (and red if there are solar prominences), with the very black occulting disk of the Moon showing in dramatic contrast. During a total solar eclipse, the sky can become sufficiently dark for the brightest stars and planets to be visible. Total solar eclipses, which last between 2 and 8 min, can only be viewed from narrow strips of the Earth's surface, a few hundred kilometers wide, on average. Total eclipses are relatively rare: about one every 18 months.

Since both the Earth and the Moon have elliptical orbits, the angular diameters of the Moon and of the Sun seen from Earth vary with time. The apparent diameter of the

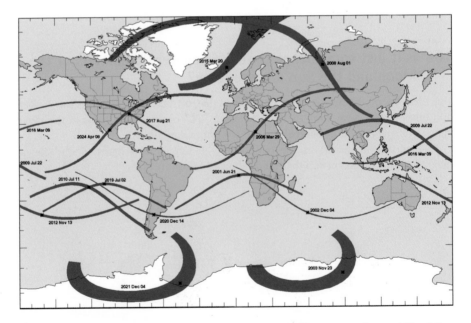

World map of total solar eclipses between 2001 and 2025. The apparent large width of the near polar eclipse totality paths is an artifact of the flat Mercator map projection. Credit: Espanak/NASA/GSFC.

Moon during an eclipse can be larger than that of the Sun (leading to longer duration of the total eclipses) or it can be smaller. In the later case, at maximum eclipse, the Moon covers only a fraction of the solar disk, which appears as a bright luminous ring. Such an eclipse is called *annular*.

When the apparent disk of the Moon never becomes concentric with that of the Sun, the eclipse is termed *partial*. Unlike total or annular eclipses, partial eclipses can be viewed from extensive regions, and this is the way total eclipses appear when viewed from outside the totality path. The amplitude of a partial eclipse is defined by the percentage of the solar disk covered by the Moon.

110 How can one tell if the Moon is waning or waxing?

Except for mariners and amateur astronomers, few people observe the sky systematically enough to notice the evolution of the Moon's appearance throughout the month. And unfortunately, fewer and fewer calendars still display the lunar phases.

How can observation of the lunar crescent, day or night, tell us whether the Moon is increasing in phase (waxing) towards the full moon or decreasing

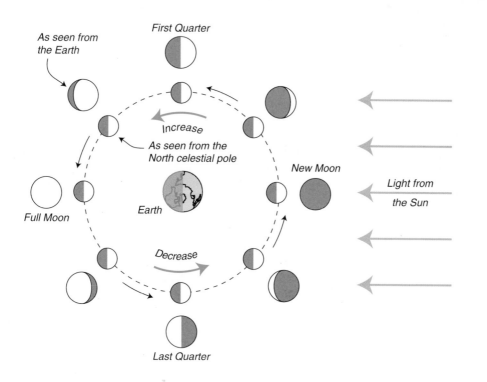

Phase	Time of moonrise (in the east)	Time of meridian passage (in the south)	Time of moonset (in the west)
New Moon	sunrise	noon	sunset
First Quarter	noon	sunset	midnight
Full Moon	sunset	midnight	sunrise
Last Quarter	midnight	sunrise	noon

(waning) towards the new moon? Readers with a talent for geometry can reconstitute the following schematic in their heads.

For the rest of us, there are two useful mnemonics (assuming that one is in the northern hemisphere). The first one is "DOC." The crescent preceeding the full moon (waxing) is in the shape of a D, then there is the full moon in the shape of an O, followed by a crescent in the shape of a C for the waning moon. The second mnemonic is "Dog comes in (the room), Cat goes out," with *coming in* = waxing, *going out* = waning. In the southern hemisphere you are "upside down" and must reverse the meaning of the cues. (If you are near the equator, the crescent can be horizontal, providing no clue at all.)

For extra help, the table below shows where the Moon appears in the sky as a function of its phase and the time of day.

111 What has been learned from our exploration of the Moon?

Had it not been for the Cold War, the Moon exploration programs of the 1960s and 1970s would not have been undertaken so early and so determinedly. Driven by military objectives and national pride, the USA and the USSR, the two superpowers of the time, launched themselves into a space race with the Moon as first prize. Although technical achievements took priority over purely scientific programs, our knowledge of the Moon (and of the solar system) benefitted enormously.

Thanks to seismic and magnetic measurements conducted in situ and to samples of lunar rocks brought back by the astronauts, the scientific contribution from the *Apollo* missions has been crucial. It has allowed for a much fuller understanding of the Moon's composition and formation, areas that had been mostly speculative until then. The complete absence of signs of life there was also established.

But the contribution of manned lunar exploration programs has not been limited to a better understanding of the Moon itself. Like the Earth, it is a very ancient body, but a geologically inert one, a frozen image of the conditions existing in the solar system during its first billion years. Our understanding of the origin and evolution of the entire solar system, of its rocky planets (Mercury, Venus, and Mars), and ultimately of our own Earth, owes an enormous debt to the men who had the courage and opportunity to walk on the surface of our shining nighttime companion.

Astronaut–geologist Harrison Schmitt picking up samples from the lunar surface during the 1972 *Apollo 17* mission (the last manned lunar landing mission). Credit: NASA AS17–145–22157.

112 How useful would it be to return to the Moon?

We already know quite a lot about the Moon. Human beings have walked on it, even driven around on it. Some 400 kg of lunar rocks have been brought back to Earth for study. We have a highly detailed map of its surface and all its major geographical features have already been named. Finally, going there is extremely expensive. So why return?

First, because we still have much to learn. From a geological viewpoint, the excursions by the *Apollo* missions surveyed only areas of easy access, the plains. Other terrains remain entirely unexplored.

Also, because the Moon could prove to be an interesting astronomical site. The far side of the Moon, which is shielded from radio interference from Earth, would be an ideal location for radio telescopes. With no atmosphere to disturb the image quality and extremely cold nighttime temperatures, it also offers many advantages for infrared observing. It may not be quite as perfect a solution as deep space for this wavelength domain (Q. 225), but, if a lunar base were to be built for any reason, astronomical research would certainly benefit from piggy-backing on those facilities.

Artist's view of a lunar factory for extracting oxygen from rocks to fuel Mars rockets.

More interesting in the long run would be to use the Moon as a training field and launch base to explore Mars and beyond. Since gravity there is six times weaker than on Earth, rockets launched from a Moon base would require significantly less propellant. The oxygen for fuel would probably have to be extracted from the lunar rocks, requiring large sophisticated infrastructures and a permanent lunar base. A permanent Moon base may seem far-fetched, but let us not forget that the Moon is really quite close, right in our backyard. If it were possible to drive there in a car, the trip would take about four months, less time than it used to take to cross the USA in a covered wagon. With a rocket, it takes only four days.

Finally, the Moon could provide one solution to our future energy supply problems. Surface bombardment by the solar wind creates helium-3 there, an isotope that does not exist on Earth. Helium-3 is an ideal fuel for thermonuclear fusion: very high yield, no pollution, and no radioactive waste. It is estimated that there are about one million tons of helium-3 on the Moon, enough to supply our energy needs for thousands of years ...

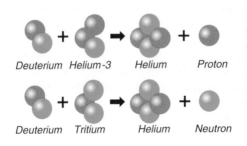

Nuclear fusion using helium-3 (top row) does not produce dangerous neutrons, as does the classical fusion technique (bottom row) using deuterium and tritium (Q. 32). The proton that is produced can be contained using electric and magnetic fields and can even be converted directly into electricity.

113 What explains the dim light suffusing the dark portion of a crescent Moon?

When the Moon is a crescent, the part of the disk that should normally be completely dark can sometimes be faintly illuminated. This is *Earthshine*, popularly known as "the old Moon in the new Moon's arms." It is the light that is reflected from Earth onto the Moon's surface.

If we were on the Moon, we would see that the Earth goes through "phases," just as the Moon does for us. But the phases are inverted with respect to each other. When the Moon is close to new Moon, the Earth, as seen from the Moon, is close to "full Earth," shining brightly. Some of this reflected light strikes the Moon, illuminating the dark portion of it, and is re-reflected back onto Earth in a pretty play of cosmic mirrors ...

Earthshine is visible for only a few days before and after a new Moon, when the crescent is thin. It disappears when there is more than a quarter Moon because the Earth is reflecting less light then and the lit part of the Moon has become dazzling.

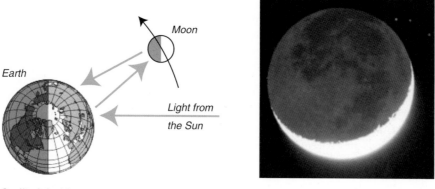

Credit: J. Lodriguss.

114 Has the *Hubble Space Telescope* been used to observe the Moon?

The *Hubble Space Telescope* was never intended to observe bright celestial objects, such as the Moon, that are easy to observe from the ground. Furthermore, since the Moon moves rapidly across the sky, it is a challenging target for the *Hubble* telescope, which is dedicated to stellar and galactic observations requiring precise but very slow tracking.[†]

Nonetheless, in 2005, *Hubble* was trained on the Moon for one very specific study. The goal was to investigate three sites for a potential lunar base, looking for sources of oxygen and other key chemicals that will be required for a permanent installation.

[†] To complete a full turn, the *Hubble* telescope needs 30 min. It could not be used to observe the Earth, which scrolls by too fast below it.

Several lunar minerals do contain some oxygen in the form of iron and titanium oxide, which can be detected by observing in the ultraviolet. This search could not be done from the ground because our atmosphere blocks most ultraviolet wavelengths.

115 "Moonstruck!" Does the Moon have an influence on human behavior?

It is easy to understand why ancient cultures imagined that the Moon had magical powers, and why its waxing and waning came to dictate so many human endeavors, when to plant and when to reap, when to marry, when to go to war. It is too spectacular an element in our sky, too big and bright, too intriguing with its changing phases, *not* to have acquired a reputation for influencing man's destiny.

"Lunacy," meaning *insanity*, comes from *luna*, the Latin word for "Moon," for the Romans believed that the mind was affected by the Moon and that "lunatics" grew ever more frenzied as the Moon increased to its full. And in the Middle Ages, folklore and superstition associated the full Moon with strange events, men going insane or being transformed into werewolves.

More recently, popular belief has it that the full Moon brings out the worst in people: more violence, more suicides, more accidents, more aggression. This supposed influence on behavior has been called "The Lunar Effect" or "The Transylvania Effect."

But what does modern science have to say?

The behavior of certain marine animals, particularly their reproductive behavior, is unquestionably synchronized with the ocean tides, which are clearly linked to the respective positions of the Moon and the Sun: think of the grunions spawning on California beaches or the horseshoe crabs on the shores of Delaware Bay. But where human beings are concerned, recent studies exploring the relationship between a full Moon and human behavior have, on the whole, come up with negative results [19]. Aggression does not seem to increase. Retirees are not more agitated. Suicides and admissions to psychiatric hospitals do not increase. The number of persons visiting emergency rooms does not go up. If a few studies do seem to find correlations, they often prove to be the results of small sampling, the selection effect, anecdotal evidence, or lack of rigor and consistency in the study protocol.

But if there is no hard evidence that the Moon provokes abnormal behavior in human beings, who would deny its power to foster perfectly "normal" behavior? How many romantic hearts can deny the magic of the light of the silvery Moon?

Celestial phenomena

116 What is a shooting star?

Shooting stars are not stars at all, although they can briefly outshine the brightest of them. What looks like a falling star is simply the luminous streak left by an interplanetary dust grain as it passes through the atmosphere. Most such particles are tiny, the size of a grain of sand, but they travel very fast – tens of kilometers per second. As they streak through the atmosphere, they heat up and are vaporized. These incandescent grains are much too small to be visible, but the atoms that evaporate from their surfaces strike air molecules, causing these to produce flashes of light, and that is the bright glow that we see.

A shooting star over Hong Kong from the 1998 Leonid meteor shower. Credit: Yan On Sheung.

This all takes place in the upper atmosphere, beginning at an altitude of approximately 130 km. The grains have typically disintegrated before they reach an altitude of approximately 80 km.

The luminous phenomenon is called a *meteor*, Greek for "high in the sky," while the physical object that causes the glow is referred to as a *meteoroid*. The Earth is constantly bombarded by meteoroids – hundreds of millions each day – but many are too small to leave a visible streak of light. You can usually expect to see about 7 meteors per hour on a clear, moonless night.

Although the vast majority of meteoroids are minute, a larger one will occasionally survive its passage through the atmosphere and strike the Earth. Such a meteoroid is then called a *meteorite*. The meteor associated with it, especially brilliant and persisting for several seconds, is called a *fireball*.

117 What causes meteor showers?

The number of shooting stars increases significantly at certain times of the year, reaching a frequency of about one per minute. Such meteor "showers" occur when the Earth passes through clouds of dust left behind by comets (Q. 70). On approaching the Sun, their heated surfaces vaporize, leaving a trail of dusty debris behind them.

The Perseids meteor shower in 1995. Credit: NASA Ames Res. Ctr.

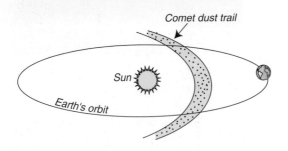

The Earth passes through several such clouds, called *swarms*, every year on its path around the Sun. The most spectacular meteor showers, named for the constellations that appear to be their points of origin, are the Perseids in August, the Leonids in November, and the Geminids in December. In a meteor shower, the glowing trails are all parallel to each other, and, due to an effect of perspective, appear to radiate out from the same point (Q. 120).

118 What causes the "northern lights?"

The northern lights, or aurorae borealis, are a magnificent spectacle of the night sky at high latitudes. They should really be called the "polar lights," however, because they also appear near the South Pole (aurorae australis). A beautiful but dangerous gift from the Sun, they are due, not to sunlight, but to the particles that make up the solar wind (Q. 47), particles with so much energy that they can damage living cells and trigger mutations.

Fortunately, we are protected from the solar wind by the Earth's magnetic field, whose "lines of force" are similar to those that can be seen around a bar magnet on a table peppered with iron filings. The magnetic field deflects the flow of particles around Earth much as a rock on a streambed deflects water in a flowing stream.

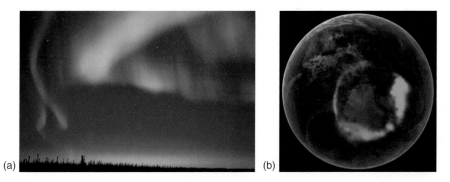

(a) Image of a spectacular aurora in Alaska. (b) A panoramic view of the southern ring of aurorae imaged by spacecraft above the South Pole. Credits: D. Hutchinson and NASA/GFSC.

Artist's view of the Earth's magnetosphere, which deflects the very damaging solar wind. Credit: NASA/GSFC.

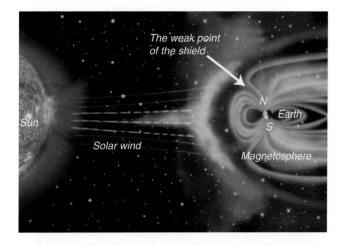

But this natural shield of ours, called the magnetosphere, has two entry points, one above each of the magnetic poles, where the lines of force re-enter the Earth. High in our upper atmosphere, at an altitude of approximately 80 km, the lines of force form a sort of magnetic funnel into which particles from the solar wind flow, collide with the atoms and molecules of the air, and excite them. These then re-release the energy they have absorbed and emit photons, i.e. light, in a process very similar to what goes on inside a neon tube. The color emitted depends on the nature of the atoms or molecules that are excited: green and red for oxygen, blue and violet nitrogen. Red aurorae are rare and indicative of particularly high-energy events.

119 What is zodiacal light?

Zodiacal light is a faint, diffuse glow that appears in the night sky in the direction of the Sun just after twilight and before the light of dawn. Clearly visible on moonless nights, it can last for several hours. It is called *zodiacal* because it is brightest along the zodiac, the belt of constellations through which the Sun, Moon, and planets appear to pass.

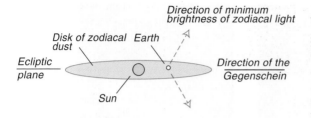

Zodiacal light, which can be as bright as the Milky Way, is caused by sunlight reflecting off interplanetary dust particles. It is easier to observe at low latitudes because it projects at a higher angle with respect to the horizon. Zodiacal light has been known since antiquity – the Persian poet Omar Khayyam called it "false dawn." Directly opposite the zodiacal light there is an analogous, much weaker glow in the sky called the *gegenschein*, from the German for "opposite glow."

Despite its low luminosity, zodiacal light is a nuisance for space telescopes, especially for infrared observations, because dust grains illuminated by the Sun radiate their own heat. Hence, observations have to be scheduled in the directions of minimum light. Fortunately, the regions of the sky affected by zodiacal light change over the year.

The gegenschein. The bright, elongated object on the left is the saturated image of Jupiter. Credit: Fukushima *et al.*

120 What causes the bright beams of light, like searchlights, that stream out from the setting Sun?

Those magnificent beams of light that sometimes emanate from the Sun near the horizon are called "crepuscular rays." They appear when light, which is being scattered through the atmosphere, is blocked in certain directions by clouds or mountains located in front of the setting Sun. The rays themselves correspond to regions of the sky that the Sun is illuminating, and which stand out brightly against the shadows cast by the clouds or mountains.

These rays are particularly noticeable when the air contains water droplets or dust particles that strongly scatter sunlight. The rays are actually parallel to each other since

Crepuscular rays. Photo: Holle, University of Illinois.

they come from the Sun which, for all practical purposes, is at infinity for us. But they appear to diverge from the Sun because of an effect of perspective, as when the rails of a train track appear to converge at the horizon.

121 Why is the setting Sun red?

Let us first talk about the color of the Sun when it is high in the sky. It is *white* then – no, not yellow as children's drawings show it. Not convinced? The Sun is too bright in full daylight for us to judge its color by looking at it directly, but we can project its image onto a sheet of paper with a magnifying glass: the image is white. The Moon, which simply reflects light from the Sun back to us, also looks white. And the same is true of clouds. Obviously, the color of sunlight for us *should* be white, the neutral color, because our eye was designed to work in sunlight. If our Sun was another type of star, a cooler one, for example, and so redder, our vision would have adapted to this other, redder, light and we would still see it as white.

When the Sun sinks low on the horizon, on the other hand, its light has to travel through a thicker slice of atmosphere. At 5° from the horizon (the thickness of three fingers held out at arm's length), it passes through 10 times as much atmosphere as when it is directly overhead, and we can look at it without being dazzled. And since the air molecules *scatter*[†] blue light more than red and yellow light, the Sun appears yellow to us. When

[†] Scattering of sunlight in the atmosphere occurs when photons encounter dust particles, water droplets, or air molecules. In the case of the air molecules, which are about the same size as the wavelength of light, the process is due to photons striking the molecules which, excited, re-emit that extra energy as new photons at the same wavelength, but in random directions instead of along their original paths. This means that many of those photons do not reach our eyes. Blue photons are energetic enough to produce this effect, but red photons, which have less energy, cannot easily penetrate the molecules, so most of them continue along their paths and do reach our eyes. The greater scatter of blue light is also the reason why the sky is blue.

it reaches the horizon, the thickness of the atmosphere is 100 times that at the zenith: the blue and green light is essentially gone and the Sun appears red.

The sky itself becomes red at sunset, especially if the atmosphere contains fine particles of dust or water droplets in suspension. These particles scatter the Sun's ruddy light and set the sky ablaze.

122 Why are sunsets usually more colorful than sunrises?

The colorfulness of a setting or rising sun is due to the scatter of sunlight by water vapor and particles in suspension in the atmosphere (Q. 121). Throughout the night, water vapor in the air tends to condense because temperatures drop. By dawn, the air has become clearer, containing less water vapor and solid particles, and the scattering of the Sun's rays is thus reduced: sunrises are not usually very colorful.

In the evening, the opposite is true. At sea, seawater has been evaporating all day due to solar radiation and increased temperatures. By evening the air has become thick with water vapor, which increases the scattering of blue light rays and reinforces the red color of the setting Sun. Sunsets over land are often even more colorful than at sea. During the day the ground warms up, which strongly agitates the atmosphere and creates updrafts that swirl pollen and dust grains upwards, increasing the number of particles in suspension.

123 What is the green flash?

According to an old legend retold by Jules Verne in his novel *The Green Flash*, it is a magic ray of light that endows those lucky enough to witness one with the power to look deeply into their own heart and recognize true love.

Well, seeing a green flash may not bring you heightened powers of self-knowledge, but its existence is not just a myth, either. It consists of a brief but unmistakably green flash of light that occurs just as the Sun sinks below the horizon. The event is relatively rare and can only occur if the air is clean and crisp and the horizon is cloudless and well defined. It is caused by atmospheric refraction, and the conditions it requires are complex and more easily met over the ocean, but the green flash can also be seen from some mountainous sites.

A beam of light coming in from a celestial object does not follow a straight path through the atmosphere. It is bent, as when a light ray passes from air to glass or water (Q. 210), because the density of air varies with altitude, creating a succession of refractions. This results in all celestial objects being slightly "raised": at the horizon, for example, the Sun's image is elevated by half a degree. The atmosphere also acts as a prism, so that the value of this elevation is related to the color of the light, being greater for short wavelengths (blue, green) than for long ones (red). The "red Sun" thus sets first, with the "green Sun" following a split second later (there is hardly any blue in sunlight at the horizon because most of it has been

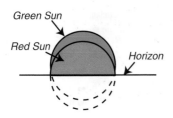

Green Sun

Red Sun

Horizon

absorbed by the atmosphere). The angular distance between the green Sun and the red one is actually too small for the naked eye to detect, only 20 arcseconds, whereas we can distinguish nothing under 1/2 arcminutes (the limit of visual resolution). So as far as our eye is concerned, the two Suns normally set at the same time.

The green flash is thus only visible when the effect is amplified somehow. The most common case is a mirage,

A fine green flash. Credit: Jim Young, JPL/NASA.

rather like what happens on a road in summer, when the sea is warmer than the ambient air. When the red Sun has already set, the tiny bit of the green Sun remaining above the horizon appears to us as if reflected in a mirror, inverted and high enough for us to distinguish it from the horizon. The green flash lasts only a second or two in the middle latitudes, and, to see it well, you always have to be positioned at least 3 to 5 m above sea level.

124 Why do we never tan in the late afternoon, even though the Sun's rays still feel hot?

The radiation we receive from the Sun is not only composed of visible light, but also of infrared rays that warm us and ultraviolet rays that tan our skin. Ultraviolet rays have a lot of energy, but they are easily absorbed by matter. Just a few millimeters of glass can stop them almost completely. You cannot get a tan through a closed window.

Air also absorbs ultraviolet rays, and most of those that come from the Sun are thus blocked by the atmosphere. The most harmful of them, called "UV-C" rays, have wavelengths of less than 280 nanometers.[†] They are completely absorbed in the very high atmosphere via the process of ionization and dissociation of nitrogen and oxygen molecules. Rays between 280 and 320 nm, called "UV-B" rays, tan the skin but are dangerous in large doses. Most of these are absorbed by the ozone layer in the upper atmosphere. As for the "UV-A" rays with a wavelength between 320 and 400 nm (i.e. just beyond visible violet light), their energy is lower still and, although they, too, can tan the skin a little, they are fairly inoffensive.

As the Sun goes down, it shines through ever thicker amounts of atmosphere, and the thickness increases very rapidly. At 30° from the horizon, its rays pass through twice as much atmosphere as when it was overhead; at 10° from the horizon, it goes through 6 times as much atmosphere and the UV radiation is enormously weakened, since absorption varies exponentially with the thickness. To see why that should be the

[†] The wavelength of visible light is between 400 (blue) and 700 (red) nanometers. A nanometer (abbreviation: nm) is one billionth of a meter (10^{-9}m).

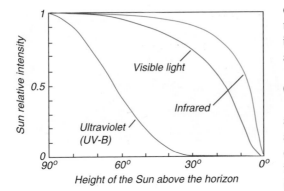

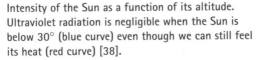

Height of the Sun above the horizon

Intensity of the Sun as a function of its altitude. Ultraviolet radiation is negligible when the Sun is below 30° (blue curve) even though we can still feel its heat (red curve) [38].

case, think about it this way: if the first meter of a material absorbs 90% of the incident radiation, the second meter will absorb 90% of the remaining 10%, so only 1% will be left. After the third meter, only 0.1% will remain.

Consequently, when the Sun drops from an altitude of 60° to 30° in the course of an afternoon, the intensity of its radiation decreases by a factor of 100. This is why it is impossible to tan late in the afternoon (or early morning), and why you cannot hope to tan in the winter in temperate latitudes even on the balmiest days. In Boston, for example, which is at a latitude of approximately 42°, the Sun rises to an altitude of 65° in June but no higher than 19° at the end of December: an hour's sunbathing in December will not even produce the effect of 2 s of sunbathing in June!

This also explains why we tend to burn so quickly in the tropics. The Sun rises high in the sky all year long and stays high for hours, so the UV rays are only slightly blocked.

As you can see in the graph, UV-B rays are almost completely absorbed when the Sun is less than 30° from the horizon. So here is the rule to follow if you want to avoid sunburn: *no matter where you are on Earth at whatever time of year, protect yourself from the Sun whenever it is higher than 30° above the horizon.* If estimating angles is not your forte, just pick up a standard-sized magazine and hold it out at arm's length, as though to read the cover. That gives you an angle of about 30°.

125 Why do stars twinkle?

Remember "twinkle, twinkle, little star?" The twinkling occurs because our atmosphere is never completely still; the air is always in movement thanks to winds and updrafts. And since the temperature of air also varies, its density – and consequently its index of refraction – varies all the time, too. It is as though a great many little lenses were shifting around above us, blown about by the winds. As a result, the light we receive from the stars looks briefly, at times, a bit more concentrated and, at other times, a bit dimmer.

The stars are so far away that we can never see their disks; they appear to us only as points of light – even through our most powerful telescopes. But each point of light appears more or less brilliant as turbulence cells in the atmosphere focus its light or cause it to diverge. The stars twinkle.

The planets, on the other hand, are relatively close to Earth and we can make out most of their disks with binoculars or a small telescope. That is not always true with the naked eye, but they can still be recognized by the fact that their light does *not* scintillate. The reason for this is that the light from different parts of a planetary disk

The planets do not twinkle because the angles they subtend are much larger than the size of a turbulence cell in the atmosphere (the angles in this drawing are greatly exaggerated).

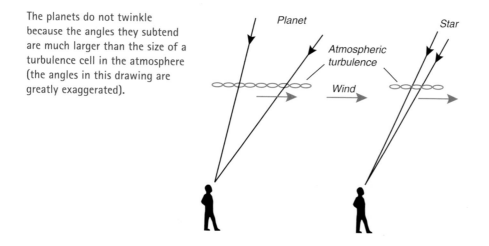

passes through many different "atmospheric lenses" before recombining in our eye.[†] What we perceive when looking at a planet is its average luminosity, and thus the fluctuations are very attenuated.

126 Why does the Moon look so large at the horizon?

The effect is often striking: close to the horizon, the Moon can look one and a half times as large as it does at the zenith. The effect is particularly strong if there are reference objects on the horizon, such as trees or buildings, but it can occur even at sea. You would be disappointed if you took a photo, however: the gigantic Moon would have shrunk. The effect is not real.

There is no doubt that the illusion results from an error our brain makes as it interprets our surroundings. The Moon looks quite normal in size if we look at it upside down (from between our legs, for example), and there is no illusion in a planetarium, either, even if silhouettes of trees or buildings appear on the artificial horizon.

Several explanations have been put forth, but none is completely satisfactory. The most promising theory is linked to our perception of the vault of the sky. The sky is not a vault – it stretches out to infinity. Still, we have the impression that there is a surface above us, and it has been proposed that we imagine it "flattened" rather than spherical, so that our brain interprets it as being closer at the zenith than at the horizon. This could be due to the way we evolved, it could be a survival factor. Since food, danger, and protection were more often to be found *around* us than *above* us, it could be that our brain evolved to favor our perception of things at ground level to the detriment

---

[†] The angular size of a turbulence cell in the atmosphere is about 1 arcsecond, whereas the apparent diameter of the planets is always larger. Jupiter can have a diameter of up to 45 arcseconds.

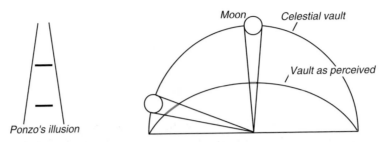

Ponzo's illusion

In the Ponzo illusion, the two horizontal line segments are equal in length, yet we perceive the top segment as being longer because our brain interprets the two converging lines as being parallel lines seen in perspective, like railroad tracks. It then draws the conclusion that the upper segment is more distant than the lower one, hence that it must be longer. Similarly, since we have the impression that the vault of the sky is flattened rather than hemispherical, the Moon may appear to be larger at the horizon because our brain interprets it as being more distant than at the zenith.

of those high in the sky [41]. In any case, convinced that the horizon is further away than the zenith, our brain automatically "corrects" our perception of the rising Moon, making it seem larger to us.

The Universe

127 How old is the Universe?

Living things obviously have finite lifetimes: they are born, live, and die. Also, we know now that planets, stars, and galaxies have finite lifetimes. The idea that the Universe had a beginning, is evolving, and may eventually come to an end is a relatively recent and quite revolutionary idea (Q. 81). The ancient Greek philosophers and most great mythologies postulated a Universe with no beginning – immutable and eternal. Even Newton, Einstein, and the astronomer Edwin Hubble, whose works underpin our current understanding of the history of the Universe, found it hard to accept the concept of cosmic evolution.

However there is no longer any doubt that the Universe "began" 13.7 billion years ago (Q. 128). How is it possible that such a number can be determined with such astonishing precision? Four different methods were used and the results of all of them are consistent.

The first method consists of calculating the age of the oldest known stars which are found in the globular clusters of our galaxy's halo. Since nuclear energy is what makes stars shine, we can determine the age of a star by measuring how much of its nuclear fuel has been used up. The oldest stars are between 8 and 13 billion years old as determined by comparison of observational data with theory. So the Universe is at least that old, and very likely a bit older if we take into account the time needed for conditions in the early Universe to become favorable for the birth of the first stars.

The second method consists of estimating the age of the oldest atoms. Radioactive elements such as uranium-235 and -238, thorium-232, and potassium-40 are unstable and disintegrate over time, and their degree of disintegration is a measure of their lifetimes. The oldest such atoms found on Earth or in meteorites are about 10 billion years old. These atoms are much heavier than hydrogen and could only have been created long after the birth of the Universe (Q. 16). Therefore, the age of the Universe must be greater than the age of the atoms in it, and so older than 10 billion years.

The Earth–Moon couple, like all objects in our solar system, is quite old: about 4.5 billion years – one third the age of the Universe. This magnificent photo of our small world was taken by the space probe Galileo on its way to Jupiter in 1990. Credit: NASA.

Globular star clusters such as the Centaurus cluster shown here, 15 000 light-years away, contain up to several million stars which are among the oldest objects in the Universe. They are over 10 billion years old.
Credit: Loke Kun Tan – Starry Scapes.

The third method is based on the time needed for galaxies and galaxy clusters to form. As gravity is a very weak force over great distances, between 10 and 40 billion years would be required for these immense structures to assemble themselves. Based on the large clusters of galaxies that seem to be still in the process of formation, it can be estimated that the age of the Universe is under 20 billion years.

The last and most precise method is based on our current measurements of the rate of the expansion of the Universe assuming the most popular model of cosmology. If we extrapolate back in time, presuming that this rate has remained constant, we come to the inevitable conclusion that the Universe began as an ultra-dense microscopic sphere, and that it cannot be much older than 13 or 14 billion years, which is, by the way, the most convincing proof that the Universe had a beginning. An even more precise value for the age of the Universe can be obtained by combining the value of the current rate of expansion with the cosmological parameters deduced from the characteristics of the cosmic microwave background (Q. 133). This provides an age of 13.7 billion years, with a precision of better than 2%.

128 How did the Universe begin?

The origin of the world is a central theme of the mythologies and cosmogonies of all great civilizations. With the twentieth century, the study of the origin and the evolution of the Universe has become part of the scientific domain. Currently, the Big Bang theory provides the most complete description. After several decades of skepticism and of trial and error, this theory has become the pillar of contemporary astrophysics, supported by a wide range of observations.

According to this theory, the history of the Universe began 13.7 billion years ago with the sudden burgeoning of a microscopic bubble of colossal energy density in space-time.[†] The young Universe was a soup of intense electromagnetic radiation and of very

---

[†] Energy density is the amount of energy in a specific region of space, in this case the whole Universe. Some of the energy is locked in radiation and some in matter. In the earliest time, the energy density is dominated by radiation, and then later matter becomes more important.

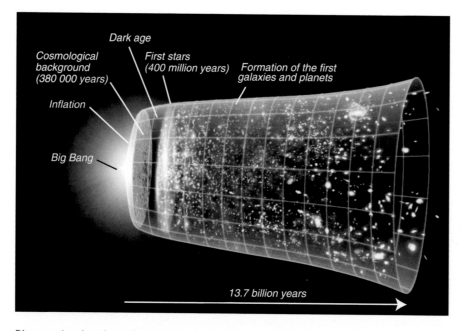

Dark age

Cosmological
background
(380 000 years)

First stars
(400 million years)

Formation of the first
galaxies and planets

Inflation

Big Bang

13.7 billion years

Diagram showing the main phases in the evolution of the Universe: the Big Bang 13.7 billion years ago, quickly followed by the period of inflation, then a slower expansion that began accelerating about 4 billion years ago. Credit: NASA/WMAP.

high-energy particles. Everything was at an unthinkable temperature and density. As the Universe expanded, it cooled rapidly. In a brief instant – some incredibly small fraction of a second – the Universe increased its volume by a factor of 10^{60} (1 followed by 60 zeros): this was the period called *inflation* (Q. 134).

Just as an ordinary gas cools as it expands, the Universe rapidly cooled during its inflation and continuing expansion, precipitating crucial transformation. The properties of the Universe changed completely over periods of nanoseconds or less. Some fundamental transformations took place in the very early phases. The fundamental forces (nuclear force, electromagnetic force, weak interaction, and gravitation) were initially combined, and then broke apart and became distinct forces, gravity being the first force to separate (Q. 144). The primordial constituents of matter such as electrons, quarks, and neutrinos were created. More complex structures took form. At the end of the first 10 minutes, the neutrons merged with the protons to form hydrogen, helium, and deuterium: hydrogen (as protons) was, as it is today, the most abundant element. Most of present-day helium (two protons and two neutrons), the gas that is used in birthday balloons, was formed at that epoch. From the point of view of energy density, radiation still dominated over matter. The atoms were still ionized and the still-free electrons were like a multitude of small mirrors that reflected light from everywhere. Space was then as bright as the surface of the Sun.

After 380 000 years of growth, the Universe had cooled significantly, down to a temperature of only a few thousand kelvins. The kinetic energy of particles had also diminished, and the electrons could combine with protons in a stable configuration

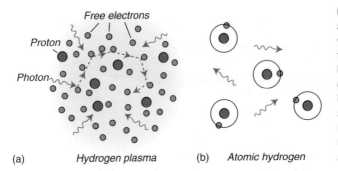

(a) In the primordial soup just after the Big Bang, collisions with free electrons prevented the light from emerging. The Universe was opaque. As it cooled down, the free electrons bonded to the protons to form atoms, opening spaces for the light to pass through. (b) The Universe became transparent – and dark ...

to create electrically neutral (not ionized) atoms for the first time. With the electrons locked "in orbit" around the protons, the "mirrors" effect disappeared and suddenly the Universe became transparent; aside from a weak infrared glow, it was dark for the first time. The cosmic "dark ages" had begun. Matter finally dominated radiation in terms of energy density. However, there were still no stars, no galaxies. Slowly, matter crowded in gigantic lumps that stretched in immense cosmic streams giving the current large structures of galaxies and clusters of galaxies. The first stars would soon form (Q. 135). The cosmic "dark age" was to last for 200–300 million years.

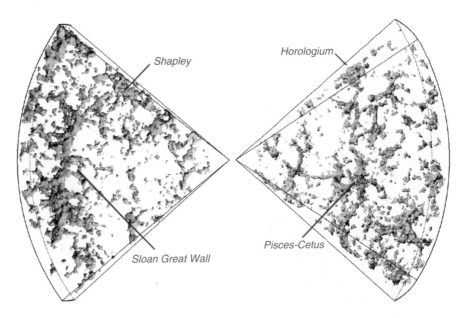

Galaxies are assembled in clusters of galaxies, the clusters in super clusters, and these into gigantic sheets and webs. These structures surround "voids," i.e. large cosmic bubbles containing few galaxies. The figure shows a sample of one such large structure extracted from a slice of the Universe near us – "Sloan's Great Wall" is more than a billion light-years long. Credit: Anglo Australian Observatory.

129 How do we know that the Universe is expanding?

Until the 1920s, almost everyone thought that the Milky Way was all there was to the Universe. Of course, there were some who surmised that the perplexing spiral nebulae were outside our galaxy, but since there was no way to measure their distances, few were persuaded. The Universe was also perceived then as being unchanging over time, a view reinforced by the small velocities of stars compared to the speed of light.

We know since the late 1920s that the Universe is expanding. The American astronomer Vesto M. Slipher (1875–1969) and the Swedish astronomer Bertild Linblad (1895–1965) had independently found solid evidence of the systematic recession of distant galaxies as early as 1914. However, it is the American astronomer Edwin Powell Hubble who showed in a study published in 1929, and confirmed later in 1930 by his colleague Milton L. Humason, that the recession speeds were increasing with the distance to the observed galaxies.

The phenomenon of the expansion of the Universe was rapidly accepted. The space-time where we live is expanding, carrying all the galaxies away from each other, like the ice floes drifting on the ocean. In whatever directions one looks, everything appears to flee from us. If we were elsewhere in the Universe or at another time of its history, we would observe the same phenomenon of systematic recession.

Although Einstein initially ignored it, the expansion of the Universe is one of the predictions of general relativity. The Russian mathematician Alexander Friedmann, in 1924, and the Belgian physicist George Lemaître, in 1927, demonstrated it to the surprise of Einstein, who declared later that this absent-mindedness was the greatest blunder of his career. Dropping the static solutions of the equations of relativity,

(a)

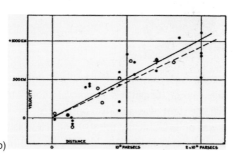

(b)

(a) Edwin Hubble (1889–1953) preparing to observe from the prime focus cage of the large Mount Palomar telescope (photo circa 1950) and (b) a facsimile of the diagram from his famous 1929 paper showing that the recession velocities of galaxies (on the y-axis) increase proportionally with distance (on the x-axis). The slope of the straight line corresponds to the Hubble constant, H_0 (Q. 130).

Friedmann and Lemaître had chosen the "dynamic" solutions. These solutions result in a space that varies as a function of time as either a contraction or as an expansion. Soon afterwards, the observations of Hubble and Humason proved that expansion was the right solution.

130 How fast is the Universe expanding?

The rate of expansion of the Universe is established by measuring the apparent receding speeds of distant galaxies from their redshifted spectrum (Q. 141), relative to their measured distances using Cepheid variable stars (Q. 18). Edwin Hubble had shown that the recession speed is proportional to the distance (Q. 129), and the value of the coefficient of proportionality, the Hubble constant, represents the rate of expansion of the Universe. It is usually expressed in kilometers per second per megaparsec – 1 megaparsec (Mpc) is about 3.26 million LY (Q. 7). For example, galaxies at 10 Mpc appear to recede at 700 km/s, those at 100 Mpc, at 7000 km/s. The value of the Hubble constant is $H_0 = 70$ (km/s)/Mpc, or about 20 (cm/s)/LY. The inverse of the Hubble constant $(1/H_0)$ gives approximately the age of the Universe.

The value of the rate of expansion was a very controversial topic during almost half a century. The determination of the Hubble constant with a precision of a few (km/s)/Mpc was one of the greatest scientific achievements of the Hubble Space Telescope – thousands of observing hours were devoted to this key project. The debate on the Hubble constant is now closed, or almost.

131 Who invented the term "Big Bang?"

Sir Fred Hoyle (1915–2001), one of the most innovative theoreticians of the twentieth century. He played a key role in understanding the formation of elements in stars and the chemical evolution of the Universe. His exclusion from the 1983 Nobel Prize in physics is considered an injustice. Hoyle was an extraordinary speaker and a prolific writer, in particular of science fiction and theater.

The term was coined by the brilliant British cosmologist Fred Hoyle, who used it to deride George Lemaître's "primeval atom" model. Hoyle opposed Lemaître's model which, he felt, was poorly constructed, contradicted some important observations, and was philosophically unsatisfying. He was also uncomfortable with the overzealous support for that model. He, himself, favored the now-discredited *steady-state theory* in which matter, instead of coming into existence all at once, is *continuously* created as the Universe expands.[†]

[†] The required rate of creation for new matter is exceedingly small: 1 atom per cubic meter every 5 billion years. The steady-state theory was supported by many cosmologists until 1965, when the discovery of the cosmic microwave background radiation essentially demolished it (Q. 133).

The Belgian priest Georges Lemaître (1894–1966), reconciling his religious and scientific vocations, pursued studies in engineering, physics, and mathematics. During the 1920s he became interested in contemporary astrophysical problems and demonstrated, to the surprise of Einstein and others, that the equations of general relativity were leading to the conclusion that the Universe was in expansion. In 1927, he proposed his "hypothesis of the primeval atom," which later became known as the Big Bang theory.

Hoyle first used the expression during one of his famous radio interviews on the BBC, which took place on March 28, 1949, and the transcript was published soon after in the magazine *The Listener* [3]. Then, years later, during a meeting in Pasadena (California) in 1960, Hoyle jocularly greeted Lemaître with: "Here comes the Big Bang man" [29]. Although Hoyle opposed the Big Bang concept until his death, he played an important but unintended role in the popularization of both the term and the theory, the term being adopted later by theoretician George Gamow and many others. Now generally accepted, it is the umbrella term for various theories covering the origin of the Universe.

132 Does the Universe have a center?

No. The Universe is expanding and the expansion is taking place everywhere and in all directions. The "center" of the Universe is everywhere: no matter where we might position ourselves in the Universe, we would see the rest of it growing ever more distant, the galaxies rushing away from us.

This is difficult to comprehend because the Universe has four dimensions and the expansion takes place in all these dimensions (including the time dimension). The analogy of an expanding balloon can help as proposed by the British astronomer Eddington in the 1930s. Imagine yourself as an ant on the surface of a giant balloon. You can move along the surface, and for you the world is represented by two spatial dimensions. Our familiar three-dimensional world has here only two dimensions; increasing the size of the balloon is like moving in time into the future.

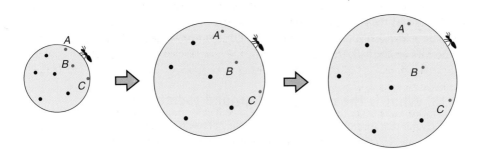

This deep field is about 5.5 · 5.5 arcminutes in angular area, that is, 36 times smaller than the apparent area of the full Moon in the sky. It is representative of what the Universe looks like in every direction. Similar observations show that the view is independent of the direction we observe, confirming that the Universe does not have a center. There are tens of thousands of objects in this image and almost all are galaxies as far away as hundreds of millions, even billions of light-years. Credit: Hook/GMOS/Gemini Obs.

The figure shows points on a balloon as a two-dimensional analogy to the Universe. As the balloon is inflated (i.e. as time advances), the dots at A, B, and C around recede from each other. An ant on the surface sees expansion in every direction and no one ant can be at the center of this expansion. Every position on the balloon is equivalent and the expansion is taking place in the same way everywhere on the balloon: there is no center on the balloon's surface.

133 What is the cosmic background radiation?

The cosmic background radiation (or cosmic microwave background) was discovered accidentally by the Canadian astrophysicist Andrew McKellar in 1940. Observing cyanogen molecules (the radical molecule CN) in the interstellar clouds of our Milky Way, he derived their temperature. Intriguingly, the temperature was the same everywhere, that is, approximately 2.4 K. This is very cold, but it is still above the point of absolute zero. McKellar did not propose a mechanism causing the excitation of the molecules, but he correctly commented upon it as being the "temperature of space," maintained above absolute zero by a mysterious origin.

Ralph Alpher (1921–2007), left, and George Gamov (1904–68).

In 1948, the physicists Alpher, Bethe, and Gamow published a milestone paper[†] regarding the formation of elements in the very young Universe, when it was very dense and very hot. They concluded that the Big Bang must have left fossil heat, like the ashes of a fire extinguished long ago. They predicted the existence of remnant light filling the Universe, corresponding to a black body at a temperature of 5 K. Curiously, no one bothered to try to observe it at the time and no connection was made to McKellar's earlier observations. This radiation was discovered by chance 17 years later by the American radio astronomers Penzias and Wilson as uniform radio "background noise" filling the sky. They won the Physics Nobel Prize in 1978 for their discovery and interpretation.

This glow, or "fossil heat," arises from the time the Universe was about 380 000 years old, when matter transitioned from opaque to transparent to light everywhere.

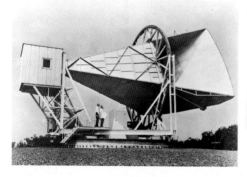

Arno Penzias   Robert Wilson

Arno Penzias and Robert Wilson discovered the cosmic microwave background in 1965. The two radio astronomers were working at the Bell Laboratories in Murray Hill (New Jersey) and were carrying out an observing program on the use of radio waves for satellite communications using a very sensitive horn-shaped antenna. The antenna is now listed as a National Historic Landmark.

[†] A paper now called $\alpha\beta\gamma$. Gamov included his eminent physicist friend Hans Bethe as second author as a joke, so that the authors' initials would be the first three letters of the Greek alphabet, but Bethe did not really contribute to it other than in subsequent discussions.

Map of the cosmic microwave background obtained in 2008 by WMAP to a greater sensitivity than COBE. The various colors correspond to temperature differences of about 0.00001 K. Credit: NASA/WMAP.

The cosmic microwave background is the memorized printout of the last reflection of primordial light when the electrons were still free and not bound to atoms. Today, after more than 13 billion years of expansion, this light has cooled enormously with a maximum intensity now at radio wavelengths in the centimeter and millimeter range.

The *cosmic microwave background* glow, or "fossil heat," hinted at by McKellar, predicted by Alpher and Gamov, and accidentally detected by Penzias and Wilson, dates from the time when the Universe was only about 380 000 years old. Matter had transitioned from opaque to transparent. The cosmic microwave background is the flash of primordial light emitted just as the free electrons bonded with the protons (Q. 128). Today, after more than 13 billion years of expansion, this light cooled so that its maximum intensity is now at radio wavelengths (7 cm) corresponding to the radiation of a black body at 2.7 K.

Very precise measurements obtained with the space missions COBE and WMAP (Q. 223) have shown that the cosmic microwave background is very isotropic. It is the same intensity in all directions in the sky, and its temperature is extremely well defined as that of a perfect black body at 2.725 K. However, one finds very small fluctuations of the order of 0.000 01 K across the sky. Despite their small amplitudes, these fluctuations represent the imprint of the irregularities in the distribution of matter at the time of the last interaction between the primordial light and the free electrons, before combining with protons to form hydrogen. The irregularities have evolved into the large structures of matter and clusters of galaxies of the Universe of today.

The analysis of the cosmic microwave background and of its spatial fluctuations also allows us to calculate the age of the Universe and to derive the relative proportions of dark energy, baryonic matter, and dark matter (Qs. 147 and 148).

134 What is cosmic inflation?

The cosmic microwave background is almost perfectly homogeneous: the differences in temperature from one point to another are smaller than 1 part in 100 000 (Q. 133). Yet these tiny differences actually are the origin of the large-scale structures, or chains of clusters of galaxies that we observe today. The fact that the temperature differences in the fossil radiation are extremely tiny is an enigma. Some 380 000 years after the Big Bang, when the cosmic radiation background was formed, the size of the Universe was much greater than the distance light could have traveled since the beginning. The different regions of the Universe could not "communicate" between themselves, forbidding energy exchange and thus preventing temperature from becoming uniform. Therefore, we would expect variations in the cosmic microwave background to be larger than detected. Even going back in time and assuming the current expansion rate, the problem becomes only worse. If we consider the time very near to the Big Bang where the laws of physics are still valid, at 10^{-43} s (one "Planck time," see Q. 138) after the initial instant, the dimension of the Universe was about 1 cm. This is much more than the distance that light or any form

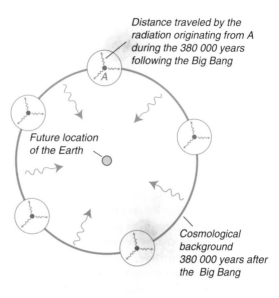

Distance traveled by the radiation originating from A during the 380 000 years following the Big Bang

Future location of the Earth

Cosmological background 380 000 years after the Big Bang

of radiation could have traveled during this infinitely small time period. Nothing in the Big Bang theory forbids homogeneity of temperature as an initial condition, but this would require a strange coincidence.

This difficulty of the original Big Bang theory, called the problem of "horizon," was solved in 1980 by a young American physicist, Alan Guth. He proposed that at about 10^{-35} s after the Big Bang, the Universe underwent a dazzling expansion, each of its elements doubling in size every 10^{-35} s, going from the dimension of a proton to that of a grapefruit at its fiftieth doubling. He called this phase of exponential expansion, *inflation*, to distinguish it from its later regular expansion as observed today. The speed of inflation would have been much larger than the speed of light, but this is not in contradiction with the principle that nothing moves faster than light, because it was not matter that was moving, but space that was growing. With the inflation model, the cosmic radiation background at 380 000 years would originate from a portion of the initial Universe that is much smaller than in the original Big Bang theory. In this model, the region was small enough so that radiation had no problem traveling across and homogenizing temperature across that miniscule volume.

Alan Guth.

The radius of the Universe observed today is not a physical boundary, but is defined to be equal to the speed of light multiplied by the time since the Big Bang, resulting in a size of the Universe of 13.7 billion LY. Of course, there is no reason to think that the Universe is limited to the observable portion. An implication of inflation theory is that the volume of the observable Universe is well within the volume of the expanding Universe (Q. 142).

The modified Big Bang theory of inflation is now accepted by most experts, because it also explains another coincidence linked to the theory of relativity, the apparent "flatness" of the Universe; that is, the radius of curvature of space is today still much

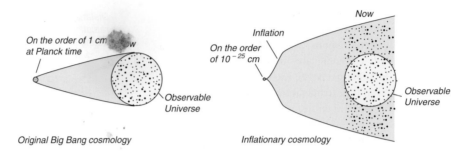

Original Big Bang cosmology   Inflationary cosmology

larger than the size of the observable Universe. This can be understood using the analogy of the surface of a balloon representing a world of two dimensions (Q. 132): inflation has inflated the balloon so much that its curvature appears now flat when only a tiny portion of its physical dimension is examined.

The mechanism driving inflation remains unknown. One proposal is that it could be due to a particle, the *inflaton* (spelled as inflation but without the second i), which would have existed in an excited state in the pre-inflationary Universe. Subject to quantum fluctuations, the inflaton field would have cascaded to a lower energy level, converting itself as follows:

$$\text{Inflaton} \implies \text{matter} + \text{radiation} + \text{cosmic inflation}$$

Imagine inflation as a "phase change." An analogy is water being transformed into ice and increasing in volume as heat is released (the process of freezing):

$$\text{Liquid water} \implies \text{ice} + \text{heat released} + \text{increase in volume}$$

Obviously, the phase change that drove cosmic inflation is fundamentally very different and involves unimaginable energy densities. It does make one wonder, though, if there is any link between inflation in the early Universe and the dark energy that seems to be accelerating the expansion of the Universe today (Q. 147). One only knows that dark energy is 10^{27} times weaker than the mechanism that was the engine of cosmic inflation. In considering observable phenomena in the Universe, we mostly ignore a possible relation between these mysterious mechanisms that affect dramatically the rate at which space is created and grows.

135 When did the first stars form?

Stars that are 10 to 12 billion years old are found in the halo of our galaxy or in globular clusters. Even these oldest stars, formed when galaxies assembled many billion years ago, show a small but significant content of heavy elements. These observations indicate that the first galaxies and their oldest stars formed from intergalactic gas already enriched with "metals."[†] From the data and models of large-scale structure and

---

[†] Quite differently from the general and chemical acceptance of the term, "metal" in astronomy refers to all elements other than hydrogen and helium.

evolution, one surmises the existence of a first generation of very massive stars with very short lifetimes of only a few hundred thousand years. Their demise produced the great fireworks that ended the "dark age" of the early Universe. These primeval stars, which formed before any galaxy, are gone forever.

The scenario of the birth of the first stellar generation is as follows. About 200–300 million years after the Big Bang, a few pockets of matter formed into lumps under self-gravitation. The first stars formed and the absence of heavy elements allowed concentrations of large quantities of gas, making it possible for stars hundreds of times more massive than the Sun to form. Despite their short lifetimes, these stars synthesized many

Artist's impression of swirling clouds of hydrogen and helium gases illuminated by the first starlight to shine in the Universe. Credit: D.A. Aguilar/CfA.

elements of the periodic table that were ejected during gigantic explosions. Nothing is left from these hyper-stars except a weak concentration of elements heavier than helium. When the Universe was a billion years old, the first protogalaxies already contained huge masses of gas lightly polluted by the heavy elements cooked in this first generation of stars.

136 How did the first galaxies form?

Two proposed scenarios attempt to answer this question. The first scenario of galaxy formation is the "hierarchical model." Clouds of intergalactic gas first collapse to form stars in small groups as mini protogalaxies. Over time, these smaller units grow and merge with other condensations to form galaxies of greater mass. In this model, the number of small galaxies remains roughly constant or increases very little with time, while the number of mid-size and massive galaxies grows considerably as the merging process continues.

The second scenario is the "downsizing model." The most massive galaxies formed most rapidly, followed by intermediate mass objects, and then more slowly by the

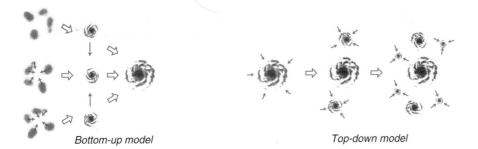

Bottom-up model   *Top-down model*

galaxies with small masses. In this scenario, it is the number of large massive galaxies that remains relatively constant or increases slowly with time. The small galaxies form slowly and their number, originally small, is constantly increasing.

The most recent observations imply that both mechanisms are at play and one or the other dominates, depending on the epoch or the environment. In the distant Universe, we observe large galaxies mixed with numerous small irregular galaxies, as well as manifestations of interaction and merging. These observations favor the hierarchical model, of growth by lumping smaller pieces together. However, in some regions, we also observe an astonishing number of massive galaxies early in the history of the Universe [21], indicating that downsizing is prevalent. How, where, and why one or the other of these scenarios dominates remains a hotly debated question.

137 Which came first, stars or galaxies?

There is still no clear answer to this question, but, according to one likely scenario, the formation and assembly of galaxies took place a few hundred million years after the

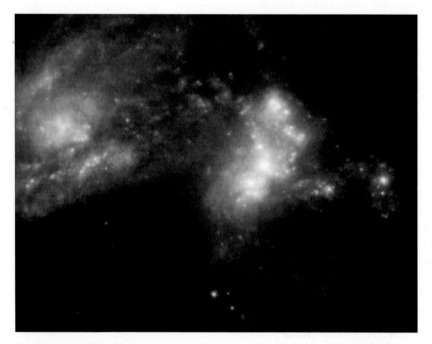

Arp 299, 135 million light-years away, is a group of at least three galaxies in the process of merging. Such collisions not only let galaxies grow and evolve, merging events also trigger episodes of intense star formation. Galaxy mergings were much more frequent in the early Universe and contributed to the formation of the Galaxies and galaxy clusters that we see today. Credit: Gemini Obs.

Big Bang, and continued until about half the current age of the Universe. Galaxies continue to form and merge today, but at a much slower rate than initially.

Recall that the first generation of stars were small in number, very massive, formed well before the first galaxies, and quickly and dramatically disappeared (Q. 135). Only when the protogalaxies were in an advanced stage of assembly and coalesced did stars appear in large numbers. That next generation of early stars and the first galaxies were nearly contemporaneous in formation. The process of initial assembly was rapid (a few hundred million years), especially for massive galaxies, and was very intense with a period where the rate of star formation was a thousand times more active than in our Milky Way today. The least massive of these first stars still exist. They can be recognized by their very low content of heavy elements (Q. 12).

138 What was there before the Big Bang?

This question fascinates many and puzzles all. For the physicist, any attempt to answer it collides head on with the fact that our current theories simply do not apply. Indeed, our physical laws cannot go back infinitely in time: they stopped at "Planck time," 10^{-43} s after the Big Bang. This is the smallest duration of time that quantum physics allows to describe. This Planck limit can be viewed as an absolute barrier, impossible to penetrate. The Universe is explainable with our physical laws only after this instant. To search what was before Planck time is like searching for a period before time itself existed, or looking for space outside space. Neither time nor space existed before the Big Bang.

Moreover, our physics is unable to describe phenomena of temperature greater than 10^{32} K (limit called "Planck temperature"). At these enormous energies, quantum and relativistic effects start to interplay and cloud the issue. Even the concepts of space and time lose their meaning. Any calculation done within the framework of current theories remains impossible.

Still, we can persist in asking: what was there before? We do not have (yet) the tools to answer. We can only speculate, using string theory to bypass the issue and to push back the limits (Q. 139). Our Universe would be only one of the realizations of a vast cosmic landscape covered with a multitude of disjointed universes, each with its own space-time. This "multiverse" or "megaverse," in the words of Leonard Susskind, one of the creators of string theory, is eternal and had no beginning. It was from this unimaginably vast and eternal generative landscape that our own Universe was born. It would be unique by its physical laws, each universe having its own set of "fundamental constants," with values very different from ours.

139 What is string theory?

String theory provides an alternative view to how particles in the Universe interact and how physical forces are manifest. In the "standard model" of particle physics based on quantum field theory, particles are considered as points. Each particle can be positioned in time and space on a relatively simple diagram with the possibility of going back in time (which is allowed for antiparticles). The different interactions between particles (and antiparticles) are determined by their mass, electrical charge, spin, and "color." The spin is somewhat related to particle rotation, and the color is some sort of charge associated with particles under the strong interaction (Q. 144). The standard model can explain the phenomena of three of the four fundamental forces: electromagnetism, weak interaction, and strong interaction, but not of gravity. Einstein's theory of gravity applies to large masses and systems in the Universe, for which there is no need to invoke the four fundamental forces at the same time.

However, in some environments the four forces must be considered together: in the environments of black holes, in the Universe just as it emerged from the Big Bang, and in aspects of the Universe when studied as a whole. Unfortunately, the formulations of quantum theory and relativity remain incompatible in several ways. This is a bit like trying to cook while combining ordinary ingredients and explosives. String theory was invented in the 1980s to circumvent those incompatibilities. Under its various expressions, it proposes a unified framework for the four fundamental forces while explaining our Universe, its origin, and its evolution.

In string theory, instead of having a number of different, point-like elementary particles, the fundamental units are reduced to a single type of object, a string. Like ordinary musical strings, these vibrate, and their multiple vibration nodes – each of which can be viewed as an elementary particle – explain the different properties of matter and energy. String theory manages to reconcile gravity and quantum mechanics by modifying the properties of general relativity in the domain of the very small, at scales on the order of the Planck length (10^{-33} cm). It then becomes possible to describe the behavior of subatomic matter in extremely strong gravitational fields and to predict interesting new phenomena.

Not everyone is ready to accept string theory, however. It is unverifiable, and the new phenomena that it predicts are not observable. To work best, string theory requires the addition of nine, ten, or even more spatial dimensions to the three physical ones (plus time) with which we are familiar (Q. 160). For some, that is simply too far-fetched. Nonetheless, string theory is currently the best tool we have to explore themes forbidden by classical physics, such as what existed before the Big Bang or why the density of vacuum energy is so small in today's Universe (Qs. 138 and 147). Elegant, but hard to test, it lines up proponents and opponents, and is subject to intense debate, but that is the beauty of science.

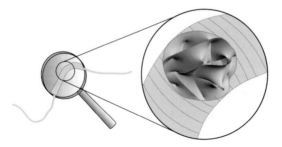

To the naked eye, a hair looks like a line: its single dimension is length. You need a magnifier to see its thickness. Similarly, space-time may appear to have only four dimensions, but may have up to 11 when viewed at the scale of fundamental particles. After S. Hawking [25].

140 If the Universe is expanding, are we also expanding?

Throughout the visible Universe the "fabric" of space-time is expanding. The expansion of space-time affects the Universe on scales of millions of light-years. This expansion is taking place at all points of space and does not have a center. The galaxies are carried away like seed pods in a river flow, hence the name "Hubble flow." However, the expansion does not affect small-scale systems like the Earth, the solar system, or even the Milky Way. At these scales, gravitation dominates and prevents the dilation of space-time. At smaller scales, atomic and molecular forces in familiar objects are even stronger, and object sizes do not change.

To visualize this, remember the analogy of the balloon being inflated (Q. 132). If the dots are actually printed on the elastic surface, they will increase in size as the membrane is stretched. But if, instead of printed dots, we have sequins or dimes glued to the membrane, these would simply move away from each other without changing size.

Even without the effects of the various cohesive forces that work at small scales, the expansion of the Universe is much too slow to be observed around us. Compare it with the compound interest rate that makes a savings account grow. With an annual rate of 10%, an investment doubles in 7 years. For the Universe, the rate of growth is enormously smaller. If, in fact, our bodies did undergo expansion, we would need half the age of the Universe, or 7 billion years, to double in size.

141 What explains the redshift of light?

The Doppler (or Doppler-Fizeau) effect is a familiar one: when a train passes by us blowing its whistle, the sound is perceived as changing in pitch from high to low. The phenomenon applies to all types of wave (light, sound, waves on water). When the emitting source of the wave is approaching, the perceived frequency of the

Waves produced by a source that is moving to the left. The wavelength appears shortened as seen from the left (bluer for light) and lengthened as seen from the right (redder for light).

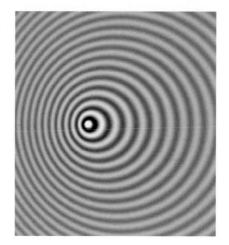

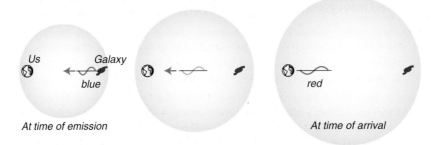

As space-time expands, the wavelength of light produced by a distant source is progressively stretched as it propagates, resulting in the cosmological redshift.

wave increases (higher pitched sound, blue-shifted light spectrum); when the source is receding, the frequency decreases (lower pitched sound, red-shifted light spectrum).

The Doppler effect appears in astronomy for all objects moving with respect to us. It applies to nearby stars that have motions relative to us (Q. 28) or to stars in orbit around black holes (Q. 145).

In cosmology, a different sort of redshift occurs. Edwin Hubble showed that distant galaxies were receding from us after measuring their speed from their redshifted spectrum (Q. 129). One might interpret this systematic spectral redshift as due to the movement of galaxies in space, as from the classical Doppler effect. This is not the case. In the expanding Universe, it is not the galaxies that are moving away from each other, but the space between them that is stretched. The wavelength redshift is caused by the stretching of the wavelength of light as it travels through space-time. The expansion is the Hubble flow (Q. 140). This is the *cosmological* redshift.

A similar effect to the cosmological redshift, that is also explained by the theory of general relativity, takes place when light grazes a massive object. The force of gravitation slows it down and the wavelength increases (is redshifted). Gravitational redshift is generally weak, but it has been observed in compact objects like white dwarfs or black holes (the redshift can be infinite, light stopping), and even in the Earth's gravitational field, using gamma rays.

142 How big is the Universe?

There are two ways to determine the size of the Universe: we can measure the distance to the furthest observable objects or derive its size theoretically.

In the first method, we derive the distance of distant objects from their red-shift, represented by the letter z (Q. 141). This cosmological redshift is due to the expansion of the Universe, that is, the stretching of space. It is a direct measure of the amount that the wavelength of light increased between the moment of its emission at the source and its capture by our telescopes. The redshift, z, is given by comparing the change in wavelength to the original wavelength through the formula

$z = (\lambda_0 - \lambda_1)/\lambda_1$, where λ_0 is the wavelength of a known spectral line identified in the distant object, and λ_1 is the wavelength of that same spectral line measured in the laboratory.

As an example, the wavelength of the ultraviolet hydrogen line, called Lyman alpha, is 121.6 nm as measured in the laboratory. Suppose that in the spectrum of a particular distant quasar, this same wavelength is observed to be 729.6 nm (in the near infrared). The redshift is thus $z = (729.6 - 121.6)$ nm/121.6 nm $= 5.0$. For $z = 5$, our cosmological model [44] produces an age of about 2 billion years when the light was emitted. There is also a simple relation between the redshift, z, of an object and the age of the Universe at the time of the production of the Lyman alpha photons. The age, t, of the Universe when the photons were produced by the quasar was simply $t_0 = t/(z + 1)$, where t_0 is the current age of the Universe, 13.7 billion years. The age of the Universe was $13.7/(5 + 1)$; that is, a little less than 2 billion years old when the distant quasar produced the photons that we receive today. This means that the light from the distant quasar has traveled for $13.7 - 2 = 11.7$ billion years to reach us, and that its "distance" is 11.7 billion LY.

The most distant objects observed have redshift of a bit less than $z = 7$. These objects are massive galaxies forming when the Universe was only about 800 million years old. The Universe was also $z + 1 = 8$ times smaller. Therefore we observe objects nearly 13 billion LY away and we have access to a corresponding volume of about 26 billion LY in diameter. We obviously cannot observe further than the distance traveled by light since the beginning of the Universe, and this defines the horizon of the observable Universe. This is a barrier that cannot be crossed.

The previous approach is limited to the measure of the observable Universe, but there is a way to tackle the problem further by using the Big Bang theory itself. In the cosmic inflation model, the Universe is much larger than its observable parts, and this by colossal factors (Q. 134). At the end of the inflationary period, the parts had moved away from each other by an unimaginable distance of about 10^{72} LY.

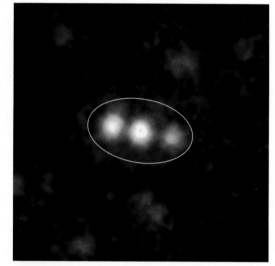

Three giant protogalaxies still in the process of formation 11 billion light-years away, as revealed by the infrared space telescope Spitzer. These three objects are among the most distant ever discovered and are close to the limit of the observable Universe. Credit: NASA/JPL–Caltech.

Since the non-observable Universe is expanding at a speed greater than that of light, we will never have access to these portions of the Universe. We ignore the true size of the Universe; however we know that it is prodigiously large and that the observable Universe is only a minute part of it.

143 Does the Universe have boundaries?

The answer to this question is yes and no! Yes, because our view into the Universe is limited to a certain distance, no because as a whole the Universe is infinite.

As the archaeologist digs to find objects that are more and more antique, the astronomer climbs up the ladder of time by looking further out into the Universe. Since the speed of light is finite, we see objects in the state they were when they emitted the light we receive from them. For example, we see light from the Sun that was emitted 8 minutes ago, 1500 years ago for the Orion Nebula, close to 3 million years ago for the Andromeda Galaxy, and 10 to 12 billion years ago for the most distant galaxies. Looking back close to the birth of the Universe, the astronomer eventually reaches an epoch before which there was no light from individual objects. This limit is the cosmic microwave background (Q. 133). For us, this is the end of the Universe, that is, our "cosmic horizon," around 13.3 billion LY away.

It is impossible to detect anything beyond that limit with the tools we now have, that is, light and electromagnetic radiation in general. One day we hope to be able to observe the primordial neutrino background filling the Universe and to get even closer to the instant of the Big Bang. However, in distance, this is a miniscule gain.

Suppose we had a craft that allowed us to explore beyond the cosmic horizon. We would certainly encounter other galaxies, because our observable Universe is probably only a minute fraction of the whole Universe (Q. 142). Would we find an edge to it? Is the Universe finite? It is not as shown in the following: imagine yourself capable of traveling at an enormous speed, much faster than that of the light, even infinite. Could you take a trajectory into the Universe that would bring you back to your departure point, proving that the Universe is finite, like the terrestrial globe? Following the predictions of the Big Bang model with inflation, and the properties of the Universe as derived from the cosmic microwave background, the answer is no. The Universe has a four-dimensional open geometry and is subject to an expansion without stops. It is without limits. The major part of the Universe is now and will remain unobservable.

144 What is the nature of gravity?

Gravity, or the force of gravitation, is one of the four fundamental forces that govern the Universe. Gravity is the weakest of the four, was studied very early, and has the most obvious effect on us. Aristotle tried to give a description of it. Galileo and Kepler analyzed its effects (Q. 194), and Isaac Newton finally provided a coherent and precise mathematical formulation for gravity: it is a force that acts at a distance and is proportional to the product of the masses involved and inversely proportional to the square of the distance between them.

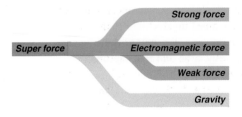

All the interactions between matter and energy in the Universe are determined by the action of one or more of the four fundamental forces. The four forces use a same mechanism: an exchange of particles, the carriers of the force, between the constituents of the Universe. In the atom, for example, the nucleus and electrons exchange photons continuously. This exchange gives rise to the electrical force that holds the atom together. Like two players exchanging a ball, the fundamental forces of nature are caught in a strange exchange through an intermediary messenger. The individual messengers – the photon, the graviton, the pion, and the W boson – are the carriers of their respective forces. It is as if the ball tied the players to one another, and acted as a binding force. In the case of the fundamental forces, the nature of the messenger defines the properties of the force and its range. At the very high energies of the early Universe, the four forces were unified. They were a single "super-force" and they are now separated and distinct. The forces and their messengers are:

- the *force of gravity* (exchange of gravitons) governs the attraction between masses and their resulting trajectories – from the movements of galaxies in collision to the path followed by a baseball;
- the *electromagnetic force* (exchange of photons) governs particles with electrical charges and the processes of emission and absorption of light and other electro-magnetic waves;
- the *strong interaction* or nuclear force (exchange of pions) ensures the stability of atomic nuclei and explains the reactions of thermonuclear fusion at the center of the Sun;
- the *weak interaction* (exchange of W bosons) governs radioactive decay.

The force of gravitation remains the most mysterious. While the origin of gravity is attributed to the exchange of gravitons, this hypothetical particle has not been detected yet.

The theory of general relativity predicts the existence of gravitational waves similar to the electromagnetic waves produced by moving charged particles. In a similar way, moving masses generate infinitesimal perturbations in the space-time fabric. These waves, called gravitational waves, are very difficult to detect, but new very sophisticated observatories have recently started operating to observe and characterize these waves (Q. 234).

145 What is a black hole?

A black hole is an object sufficiently dense and massive that its gravity prevents any light from escaping. These objects were predicted at the end of the eighteenth century by the English pastor, geologist, and amateur astronomer John Mitchell and the French mathematician Pierre-Simon Laplace (Laplace called them "dark bodies"). The concept reappeared in 1915 when German physicist Karl Schwarzschild introduced it in his interpretation of the equations of general relativity. However, one had to wait until the 1960s for a full mathematical description, in particular, by the American physicist John Archibald Wheeler, who created the term "black hole" in 1968. This concept has generated a huge interest among physicists, astronomers, and the general public as well.

The theory of relativity describes the environment of a black hole: disrupting space-time and leading to surprising phenomena, such as event horizons, singularities, slowing and freezing of clocks – many of which have been enthusiastically exploited by science fiction writers. The main characteristic of a black hole is its "event horizon" whose

Karl Schwarzschild
(1873–1916).

radius is called the "Schwarzschild radius." This dimension is given by the expression $R = 2GM/c^2$ where M is the mass of the black hole, c, the speed of light, and G, the gravitational constant. It manifests as follows: if an object of mass M is compressed to this radius, nothing can stop its further collapse to form the gravitational singularity that is the origin of the black hole. At the horizon of a black hole, time stops, as measured from our distant frame of reference. From an external observer's point of view, nothing can be detected inside the Schwarzschild radius.

The existence of black holes is not proven with absolute certainty. However, a growing number of high-energy phenomena and of objects showing bizarre behavior support their existence. For example, these include active galaxy nuclei and quasars, where the production of a colossal amount of energy could only be accounted for by the presence of black holes having masses many millions of times that of the Sun. The stupendous luminosities of quasars can be explained by the swallowing of large quantities of matter, perhaps entire stars, which are accelerated and heated when falling towards the black hole.

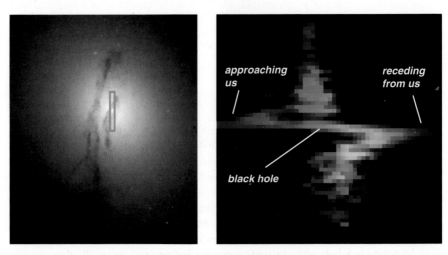

Signature of a black hole: a vortex of matter spinning at fantastic speeds. Black holes can be detected spectroscopically thanks to the Doppler effect. In this spectrum of the center of galaxy M84 (blue rectangle in left panel) obtained with the Hubble Space Telescope, matter is whirling around with a velocity of up to 400 km/s, betraying the presence of a huge invisible mass at the Galaxy's center. This black hole is estimated to be 300 million times as massive as the Sun. Credit: NASA/ESA.

Schematic diagram of the eclipsing binary system M33 X-7 in the nearby galaxy M33. A blue supergiant 70 times the mass of the Sun has a black hole of 17 solar masses in orbit around it, one of the most massive "stellar" black holes known. The orange disk represents accreted matter around the black hole, where x-ray transients are produced. In this eclipsing system, the properties of the star and of the black hole can be determined in remarkable detail. Credit: Chandra Obs./NASA.

The large orbital velocities of stars close to the centers of many galaxies can only be explained by the presence of a *non-luminous* mass of several millions of times that of the Sun confined to a volume of less than one light-year, or a black hole. Our galaxy itself, the Milky Way, appears to harbor a black hole of about four million times the mass of the Sun. This is the best interpretation for the orbital velocities of several thousand km/s of the stars that orbit close to the galactic center. These supermassive black holes originate at the time of galaxy formation.

Black holes related to stars also appear to exist. Transient and powerful x-ray bursts have been observed originating from stars in our galaxy and from neighboring galaxies. The bursts arise from gas that is captured by a companion star transformed into a black hole (Q. 16), and is violently heated, emitting x-rays while falling into the gravitational field of the black hole. "Repeating" x-ray transients are in several ways the "smoke" betraying the presence of a black hole.

146 Can anything escape from a black hole?

In classical black-hole theory, matter surrounding a black hole can only be swallowed in an irreversible way. No matter and no information can escape from a black hole, and its mass can only grow with time.

In 1974, however, the British physicist Stephen Hawking found an astonishing result while combining the principles of quantum mechanics and general relativity: a rotating black hole can "evaporate." This evaporation happens through the process of particle–antiparticle pair creation close to the black-hole horizon. Normally, these pairs of particles recombine quickly and disappear. However, near the horizon, a particle may fall back into the black hole while its partner escapes before annihilation. Particle creation takes place at the expense of the black-hole gravitational energy. The particle that escapes as radiation (photons) carries with it a minute part of the mass of the black hole. Hawking calculated the evaporation rate to be proportional to the inverse square of the black-hole mass. Therefore, this radiation, called "Hawking radiation," is very weak for black holes more massive than the Sun, but as the mass diminishes slowly, the evaporation rate accelerates in the final phase; it ends with a spectacular explosion of elementary particles.

Thus, information could escape from a black hole but it would be in an extremely rudimentary form. No final explosive phase of black holes has yet been observed.

147 What is dark energy?

Measurement of the expansion of the Universe shows that it appears to contain a repulsive force called "dark energy" that acts like antigravity. This mysterious force fascinates physicists and cosmologists, because it is a fundamental element of the nature of our Universe.

The concept of dark energy was introduced to explain the *acceleration* of the expansion of the Universe. This surprising phenomenon was discovered in 1998 by two teams of astronomers studying the behavior of the expansion of the Universe at different distances and periods of the past. They were using distant supernovae as calibrated mileposts. The supernovae result from the detonation of white dwarfs (Q. 16). While they expected a slowing down (caused by gravitation due to the intrinsic mass of the Universe), the researchers found to their great surprise, an *acceleration*. Imagine your astonishment if you saw a ball thrown up in the air accelerate away instead of falling back into your hands.

How can one explain this accelerating expansion? Some have proposed to modify Einstein's theory of general relativity. However, it is very difficult to do so without triggering serious conflicts with so many well-explained phenomena and creating more complexity. Few are ready to let go of such a robust and productive theory.

A more promising avenue is to exploit the hypothesis of the cosmological constant (the term is already in Einstein's equations) based on *vacuum energy*. One of the strangest results of quantum mechanics is that even a vacuum has energy. The vacuum, space-time without any matter, is in fact a rich and complex physical medium. Particles such as electrons, positrons, and photons are continuously created and destroyed. These particles, called virtual particles, are the products of quantum fluctuations that occur spontaneously in space. The quantum agitation of virtual particles generates an energy

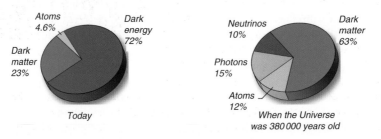

The content of today's Universe compared to 13.7 billion years ago, about 380 000 years after the Big Bang. Currently, the Universe appears to be dominated by dark energy that is driving its accelerating expansion. Note the importance of dark matter in both cases (Q. 148). Credit: NASA/WMAP.

background, "vacuum energy," that modifies the behavior of atoms. Its effect and value have been measured, and this is the energy that could be driving today's accelerating expansion of the Universe.

Unfortunately this interpretation is flawed. The value of vacuum energy should be much larger, in fact 10^{120} times (1 followed by 120 zeros) stronger than what is observed. If this was the case, the Universe would have been strongly accelerated since the beginning, preventing the formation of any of the structures that we see everywhere today: things we observe would not exist. We would not exist. In physics, a disagreement by a factor of 10 between theory and observation can deserve ridicule, or be fatal – in this case, the error is absolutely outrageous. Should we throw away vacuum energy?

String theory comes to the rescue again, merging general relativity and quantum mechanics (Q. 139). The theory offers a solution, because multiple configurations of the Universe are theoretically possible, each with its own physical laws, forces, and constants. The cosmological constant, based on vacuum energy, can take enormous values in immensely distant regions of our Universe, and very small values in others, such as the region in which we live. In our Universe (or region of the Universe), unknown particles that we hope to discover soon, would make the cosmological constant almost zero, or 10^{120} times weaker.

Nevertheless, the mystery of dark energy is not resolved. The story is to be continued! Hence, the ironical and provocative paraphrasing of theoretician Leonard Susskind of Descartes' words: "*I think, hence the cosmological constant must be very small.*"

148 If we cannot see dark matter, how do we know that it exists?

The first person to propose that there were large quantities of non-visible matter in the Universe was the Swiss astronomer Fritz Zwicky (1898–1974). In 1933, while studying a large cluster of galaxies, he noticed that its total mass was much larger than the sum of the masses of the individual galaxies forming the cluster. The result was ignored, and one had to wait 40 more years to see the evidence accepted from diligent observations made by American astronomer Vera Rubin and her colleagues. Measuring the velocities of stars orbiting the center in

Fritz Zwicky and Vera Rubin.

several galaxies, she derived their total masses, just as we determine the mass of the Sun from the parameters of the Earth's orbit (Qs. 44 and 45). To her surprise, Rubin found that the total masses were much greater than the sum of the masses of stars, gas clouds, and dust they contained. This situation was dubbed the "missing mass," but the term "dark matter" is now used. This dark substance is truly some sort of matter because it has mass, but it is matter that does not produce electromagnetic radiation nor absorb nor scatter light from its surroundings. We do not know its nature.

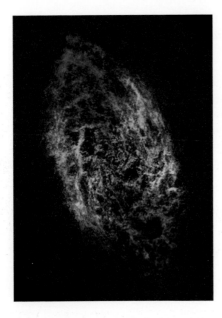

Velocity field of the neutral hydrogen gas in nearby galaxy Messier 33 (the blue color represents gas approaching us; red, gas that is receding). The orbital velocities of the stars and gas indicate the presence of much more mass than the value obtained by simply adding the masses of the visible stars, gas, and dust. There is four times as much dark matter as "visible" mass, whether observed at x-ray, ultraviolet, optical, infrared, or radio wavelengths. Credit: NRAO/AUI.

The matter with which we are familiar is made of protons, neutrons, and electrons. This family of particles is called "baryons" (from the Greek *heavy*). Baryons belong to hadrons, a larger group of elementary particles including mesons and *glueballs* in addition to baryons. Dark matter could still be baryonic, that is, made of protons, neutrons, and electrons, and solely associated with small mass objects, hence invisible in practice. The main problem is that one would need many such objects, enormously more than visible baryonic matter.

On the other hand, dark matter could be non-baryonic, for example made of neutrinos (Q. 235) of very small mass, or of strange and more massive particles such as the hypothetical *neutralinos*, with an energy equivalent to a hundred times that of the proton, or *axions* of minute energy.

Observations reveal that dark matter is about five times more abundant than the baryonic matter that we see. It is rather worrying that the major part of the matter in the Universe is in a form that we essentially ignore because we cannot see it! We expect a great deal from the new high-energy particle accelerators such as the *Large Hadron Collider* and the *International Linear Collider* to clarify this problem.

Known universe

Dark energy

Dark matter

Since there is equivalence between mass and energy, one must tally both to describe the Universe. Hence, the content of the Universe is quite intriguing: 73% in dark energy that we do not understand, 23% of non-baryonic matter of unknown origin, and only 4% of ordinary matter, that is, atoms, like those that make stars, the planets, and all the life surrounding us!

149 Were the laws of physics the same in the early Universe as they are now?

Certain dimensions, relationships, and characteristics of the Universe are so paramount that, if any one of them was different, everything would change. The question is – have the speed of light, the electron charge, Planck constant in quantum physics, and other physical constants always kept the same values? In 1928, British theoretician Paul Dirac published a paper where he juggled mathematics to illuminate some surprising relations between the various physical constants of nature. Calculating the ratio between large cosmic numbers and microscopic quantities, he was struck to find numbers of the order of 10^{39} or multiples of it – for example, the approximate number of protons in the Universe is $(10^{39})^2 = 10^{78}$. The ratio of the square of the electron

Paul Dirac (1902–84).

charge to the strength of gravitational attraction between a proton and an electron is: $e^2/m_e m_p G = 10^{39}$. This is also the value of the ratio of the size of the observable Universe to the size of a quantum particle.

These coincidences made Dirac wonder if the current state of the Universe depends on the values of the fundamental constants of nature, and consequently, that its evolution could be affected if these "constants" were not really constant. This topic has been frequently debated since then and measurements have been conducted to compare behavior of physical phenomena at the beginning of the Universe with those of today.

Such measurements, supported with laboratory experiments, show that if there have been changes, they are smaller than 10^{-13} per year; that is, less than one part in one thousand since the birth of the Universe. At that level, if the constants ever did change, one can say that their influence on the evolution of the Universe is negligible. On the other hand, the expansion of the Universe should not have had any influence on the value of the fundamental constants either.

Admittedly, we have no idea of what determines the value of the fundamental constants such as the electron mass, the speed of light, or the gravitational constant. We know their values with a dazzling precision, but we do not know why these constant have their precise values. What we know for sure is that the existence of galaxies, stars, and planets depends on a delicate balance of these values (Q. 164). For example, if the

This spectrum of light produced by a distant quasar ($z \sim 6$) when the Universe was barely one billion years old looks the same as what we would see in today's Universe (except for the large redshift). This demonstrates that the properties of the main atoms were the same then as today. Credit: Jiang et al., Gemini Obs.

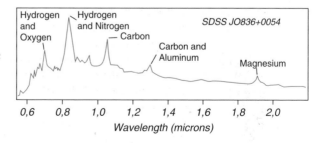

electron were more massive, the atoms and the chemistry that we know would not exist. If the force of gravitation were stronger, the Universe would be filled with black holes. String theory suggests that the fundamental constants can take very different values in parallel universes and hence lead to completely different physical laws. While fanciful, this is not testable, however.

150 How much antimatter is there in the Universe?

Antimatter is not a figment of some theoretician's wild imagination. It does exist. For each type of elementary particle (proton, neutron, electron), there is a corresponding antiparticle. For particles with electrical charges, the antiparticle has an opposite charge. The antielectron, called the *positron*, has the same mass as the electron but the opposite electrical charge. The antiproton has the same mass as the proton but a negative charge. The antineutron has no charge, but it interacts with the other particles in ways opposite to those of the neutron.

Antimatter was predicted to exist before it was discovered. In the 1920s, quantum physics was still in its infancy and was not fully developed. It could only be applied to particles moving at slow speed as compared to the speed of light. British physicist Paul Dirac solved the antimatter problem in 1928, for which he received the Nobel Prize in Physics in 1933. His relativistic equation led to two solutions, one for normal electrons, the other for an electron with a positive charge. Based on that, he made the bold prediction that antielectrons exist. Shortly afterwards, they were discovered in cosmic rays, and, since then, all the antiparticles have been detected. Normally, we do not encounter antimatter on Earth because as soon as a particle meets its antiparticle, both are annihilated, transformed into energy in the form of gamma rays.

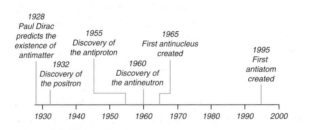

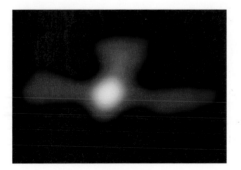

Intriguing cloud near the center of our galaxy with strong gamma emission, perhaps due to matter/antimatter being annihilated. Credit: W. Purcell *et al.*, OSSE, Compton Obs./NASA.

However, antimatter is regularly produced in particle accelerators: whenever energy is transformed into particles, the result is always an equal number of particles and antiparticles. Antimatter is the most expensive "commodity" in the world. It can be fabricated only in the most powerful accelerators and can be conserved only briefly in powerful magnetic fields. Its production cost is estimated at US $300 million per milligram.

Although there is no a priori reason why there should be more matter than antimatter, the latter constitutes just an infinitesimal fraction of the total mass of today's Universe. It is found in cosmic rays produced by supernovae, active galaxy nuclei, and interactions between interstellar clouds. Our own cosmic neighborhood is made of matter: the Sun floods us with its atoms, the solar wind (Q. 47). If these were antiatoms, their annihilation as they collided with atoms in our atmosphere would produce enough gamma rays to fry us. Our galaxy seems to be made mainly of matter (Q. 156) because it produces only weak gamma ray emissions. Also, the galactic gamma radiation is easily explained by familiar high-energy phenomena without involving antimatter. If the Galaxy contained pockets of antimatter, their interaction with neighboring regions made of matter would result in intense zones of gamma radiation. This is rarely seen.

It is likely that all the galaxies of the observable Universe are also made of matter, not antimatter. Antigalaxies constituted of antistars could, of course, exist without our noticing it – the electromagnetic waves produced by these antiobjects would be indistinguishable from the waves produced by normal objects. However, cosmologists believe on theoretical grounds that our Universe is fundamentally asymmetrical: matter dominates antimatter.

The reason for this imbalance can be traced to the very beginning of the Universe, when matter and antimatter existed in nearly equal quantities. Particles and antiparticles annihilated each other, producing gamma rays which, in turn, produced new particle–antiparticle pairs in a continuous, rapid process of annihilation/materialization. Then, as the Universe expanded, its temperature dropped, and at about 10^{-11} s after the Big Bang, when it reached 10^{13} K, the gamma ray photons no longer had enough energy to create new pairs. The annihilation/materialization reaction became irreversible with, by sheer chance, a very slight asymmetry: for every 1 000 000 000 antiparticles there were 1 000 000 001 particles. Almost all matter/antimatter was then annihilated in this final cosmic "firestorm," leaving only a small excess of baryons – and lots of photons. The eventual result was our matter-dominated Universe. In the absence of that minute initial asymmetry, the cosmos would have totally transformed itself into radiation. No matter – or antimatter – would have survived.

151 How many galaxies are there in the Universe?

Over one hundred billion – and still counting. The total number is easily extrapolated from very deep exposures made with large ground-based and space telescopes. In a recent deep field observation with the Hubble Space Telescope, 10 000 galaxies were counted over an area of the sky which is only about one hundredth the area of the full Moon on the sky (see image below). This surface density translates to about 1000 galaxies per square minute of arc.

Approximately 10 000 galaxies are visible in this Hubble Space Telescope image of a deep field made in 2003–4 (exposure time: 11.3 days). The smaller, reddish objects are among the youngest and most distant galaxies known, witnesses from a time when the Universe was only one billion years old. Credit: NASA/ESA.

Even if galaxies assemble in clusters and large-scale structures, the majority of them are isolated. The spatial distribution of galaxies is not homogeneous. Deep images drilling into the Universe out to distances of billions of light-years provide representative samples. Extrapolated to the whole sky, statistics based on the Hubble deep fields reaching as deep a magnitude as 30 (Q. 5) imply that there are about 200 billion observable galaxies, each with star populations from several tens of billions to a few hundred billions.

152 How many different types of galaxy are there?

In 1925, Edwin Hubble set up a classification of galaxies based on their visual appearance which is still used today: *elliptical galaxies* for those with a spheroidal shape, *spiral galaxies* (normal or barred), and *irregular galaxies*. About one third of all galaxies are ellipticals and most of the rest are spirals; irregulars represent only a few

Elliptical   Spiral   Barred spiral   Irregular

Examples of Edwin Hubble's classification of galaxies. From left to right, M87, M74, NGC1300, and NGC1427. Credit: CFHT for M87, NASA for the others.

percent. However, observations over recent decades have revealed a world of galaxies that is much richer and more complex than this simple classification would indicate. Galaxy sizes extend over several thousand light-years for small ones and up to more than 100 000 LY for giants.

Our galaxy is only one among billions of others (Q. 151). Some galaxies are massive giants, like the great spiral Andromeda Galaxy, our neighbor at 2.9 million light-years away and others are dwarfs, like the nearby small irregular Magellanic Clouds (Q. 155). Numbers of galaxies herd close into small groups of less than 10, or in larger clusters of several hundreds, and even in

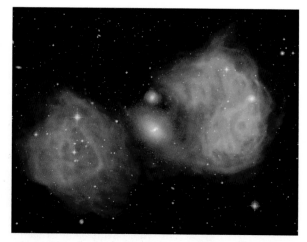

The giant elliptical galaxy NGC 1316 is at the center of two immense radio lobes (orange color) that extend over more than 600 000 LY. This image is a radio/optical composite. Credit: STScI/NRAO/AUI.

superclusters that gather several thousand galaxies in a volume of a few million light-years in diameter. The rest are spread more or less uniformly across space.

The shape and size of galaxies often depend on the wavelength used to observe them. At optical and infrared wavelengths, stellar light dominates and the distribution of the stars determines the shape. At radio wavelengths some galaxies produce a very different picture, showing small bright central nuclei that spew out gigantic jets of radio emission spanning hundreds of thousands, even millions, of light-years. These are produced by synchrotron radiation, which occurs when high-energy electrons are caught in intergalactic magnetic fields (Q. 19).

In any case we must remember that what we "see" of those galaxies is only the tip of the iceberg: most of their mass is "invisible," composed of dark matter whose nature is unknown (Q. 148).

153 What is the Milky Way?

The *Milky Way* is the name of our galaxy, the majestic stellar system of more than 100 billion stars of all ages in which our solar system is embedded. All the stars that we see with the naked eye belong to this system. Its myriad stars, together with the interstellar gas and dust it contains, are held together by the force of gravity.

The Milky Way can be seen at night as a wide, luminous band dividing the heavens into two. Its beauty must surely have been a source of wonder for our human race since we first looked curiously up at the sky. The ancient Egyptians saw it as grains of wheat strewn across the sky by the god Isis. For the Greeks and the Romans, it was the milk that the infant Hercules caused to gush from Juno's breast (the word galaxy comes from the Greek *galaktos*, milk). The Incas imagined it as a trail of gold dust, and the Ottawa Amerindians saw in it the image of a creek bed blurred by the frolicking of cosmic tortoises.

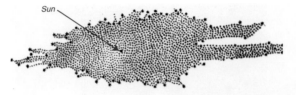

Herschel's map of the Milky Way. It is flawed because interstellar dust blocked his view of the distant stars.

Democritus, Lucretius, and other ancient philosophers surmised that this pale celestial streak was the cumulative effect of stars so distant that they could not be distinguished individually. Actual proof of that came in 1610, when Galileo turned his telescope to the Milky Way and reported: "The Milky Way is nothing but countless stars clustering in small groups. Indeed, wherever one points a telescope, an immense crowd of stars comes straightaway into view."

Harlow Shapley
(1885–1972)

In the mid-eighteenth century, Jean-Henri Lambert, Thomas Wright, and Immanuel Kant speculated that the Milky Way was a gigantic disk of stars with the Sun at its center. William Herschel confirmed that view in 1785. Having counted the number of stars visible in different directions, he found approximately the same density all along the disk and concluded that this could only be explained if our world sat at its heart.

Then, in the 1920s, the American astronomer Harlow Shapley mapped the globular clusters that surround the Galaxy and showed that the Sun and the Earth lie nowhere near the middle. This took humanity one more step away from the age-old, cherished belief that we were at the center of the Universe. A second Copernican revolution, one might say.

154 What type of galaxy is the Milky Way?

The Milky Way is a barred spiral galaxy made up of a central bulge and a swirl of spiral arms, with most of the matter – dust, gas, and stars – being contained in the arms concentrated in a flattened disk. Regions of star formation also are concentrated in the arms. The disk of the Milky Way is about 100 000 LY in diameter and about 3000 LY thick. The Sun sits about 26 000 LY (approximately half the galactic radius) from its center. Our galaxy has at least four main spiral arms and several shorter ones called *spurs*. The Sun is at the middle of one of these secondary arms, the Orion spur, and also is approximately in the middle of the disk. The Sun and its coterie of planets orbit around the center of the Milky Way at a speed of 225 km/s, meaning that it takes about 220 million years to complete one complete rotation.

It is not easy to determine the structure of our galaxy because we are seeing it from inside a disk filled with dust clouds that absorb visible light. Fortunately, observations in infrared light, which can penetrate the clouds,[†] allow us to see straight through to the central bulge. The human eye is not sensitive to infrared light and one needs

[†] The dust grains are approximately the same size as the grains of chalk dust on a blackboard eraser, i.e. half a micron (1000th of a millimeter) across. Since this is close to the wavelength of visible light, the dust grains tend to block these wavelengths, reflecting and scattering them. Infrared light, with its longer wavelength, is much less affected, and radio waves are not affected at all.

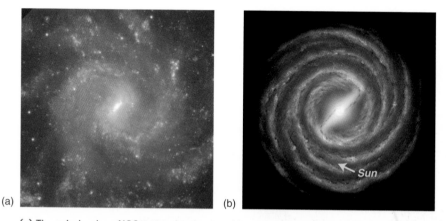

(a)   (b)

(a) The spiral galaxy NGC 7424 closely resembles our galaxy in shape, size, and content.
(b) Artist's view of our galaxy. Credit: Gemini Obs., and R. Hurst/NASA/JPL-Caltech.

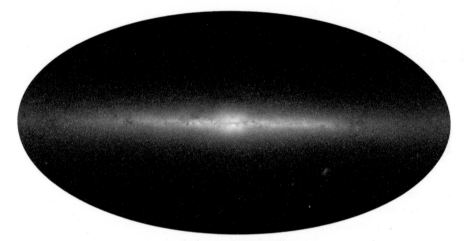

Panoramic view of our galaxy at infrared wavelengths distinguishes 500 million stars. The central bulge of the Galaxy is clearly visible. The two small whitish patches in the lower right quadrant are the Magellanic Clouds, the Milky Way's dwarf satellite galaxies (Q. 155). Credit: 2MASS survey – Univ. of Mass., Caltech, NASA and NSF.

to use special instruments for observation. For example, a few years ago the 2MASS (*Two Micron All Sky Survey*) project imaged the whole sky in the infrared using two small telescopes, one in the United States for the northern hemisphere, and the other in Chile for the southern hemisphere. The whole survey, designed to penetrate the dust to show a more detailed structure of the Milky Way, took four years to complete.

Radio astronomy provides an additional powerful tool for exploring our galaxy's structure, since radio waves are not affected by dust. Clouds of atomic and molecular hydrogen and molecules of carbon monoxide (CO), abundant in the spiral arms, produce emissions at radio wavelengths that allow us to map the position of the arms.

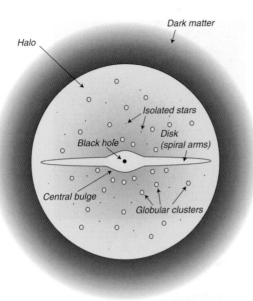

Sketch of our galaxy. The diameter of the dark matter halo permeating and enveloping is estimated to be three times as large as is represented here.

The center of the bulge, the *nucleus*, contains star clusters, dense clouds of dust and gas, and a black hole that is associated with strong radio emission. This black hole, which is of relatively modest size for a galaxy, is about four million times the mass of the Sun (Q. 145).

Our galaxy is wrapped in a large halo having a near-spherical shape that contains globular clusters as well as isolated stars. Globular clusters are compact groups of hundreds of thousand of stars crowded into spherical volumes just a few light-years across. The isolated stars and those in the clusters are very old – witnesses to the formation of our galaxy. The Galaxy and its halo of globular clusters are permeated by and enveloped in a second huge halo of dark matter (Q. 148) over 300 000 LY in diameter. The dark matter may have a mass exceeding 10 times that of the visible matter.

155 What are the Magellanic Clouds?

The Magellanic Clouds are two dwarf galaxies in orbit around our galaxy. Called the Large Magellanic Cloud and the Small Magellanic Cloud, these are small siblings of our own galaxy. They are visible to the naked eye from the southern hemisphere, and are particularly noticeable during moonless nights, where they appear as two chunks detached from the Milky Way. They are named in honor of the great Portuguese navigator Ferdinand Magellan, whose crew observed them during the first circumnavigation of the world in 1519–22. Medieval Persian astronomers had cataloged them well before that, however, as they are visible from the southern tip of the Arabian Peninsula.

The Magellanic Clouds are true galactic systems, made up of large quantities of dust, gas, and several billion stars. The Large Magellanic Cloud has a mass one tenth that of

This mosaic of the Small Magellanic Cloud, composed of 500 individual images, shows the nebulae inside it particularly well, as well as a large globular cluster (the diffuse, bluish object at the top). Credit: NOAO/CTIO.

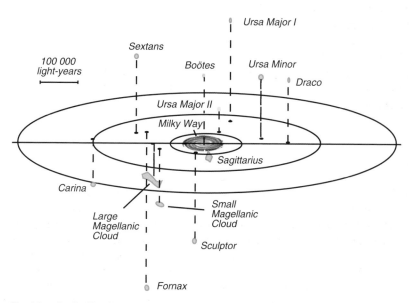

The Magellanic Clouds are not our galaxy's only satellites; there are at least a dozen of them, all dwarfs. One such, Sagittarius A, discovered in the 1990s, is already close to the center of the Milky Way and is being cannibalized by it.

of our galaxy and the Small Magellanic Cloud a mass one hundredth of it. The large amount of interstellar gas, nebulae, massive stars, and supernova remnants in both galaxies today indicates that intense star formation activity is in progress.

The Magellanic Clouds are about 150 000 LY away. Computer simulations indicate that they are currently approaching our galaxy and will merge with the Milky Way in a few tens of billions of years.

156 How does the sky appear in different wavelengths?

Gamma rays

X-rays

Visible

Infrared

Radio

Maps of the sky at different wavelengths. Credit: NASA.

Beginning with the advent of radio astronomy in the 1930s and enriched by observations from space starting in the late 1950s, we can now see the sky as never before. It is as though we had acquired eyes capable of perceiving the "colors" of radio, infrared, x-rays, and gamma rays. What does the sky look like at those wavelengths?

The images on the left are in color, but the colors used here do not represent wavelengths, they represent the *intensity* of the radiation, increasing from violet to red and, finally, to white [32].

In gamma rays, the central band is the Milky Way. Its diffuse glow is generated by the collisions of cosmic rays with the atomic nuclei of interstellar clouds. The blue spots that appear outside the plane of the Galaxy correspond to active nuclei in distant galaxies.

In x-rays, emissions by very hot gases are what cause the sky to shine. The Milky Way appears in blue due to absorption by interstellar gas. The white spots are supernova remnants, and the thin black strips are regions where no observations were made.

In visible light, we have our familiar view of the Milky Way. Several portions are obscured by dust clouds. Most of the discernable individual stars are relatively close to us (a few thousand light-years at most), while the more distant ones are either too faint to see or are concealed by dust.

In the infrared (12–100 μm wavelength), what is visible is mainly radiation from interstellar dust heated by the light of neighboring stars.

Finally, at radio wavelengths (408 MHz), we see the emission from ionized interstellar gas or hot plasmas. The large arc-shaped feature in the center is the remnant of a relatively nearby supernova which exploded several thousand years ago.

157 What is a nebula?

Nebulae (Latin for "clouds") are accumulations of interstellar gas and dust that produce light, either because their atoms are excited and fluoresce or because the microscopic dust they contain reflects light from nearby stars. Nebulae vary widely in size, from less than one to more than 1000 LY across, while their masses range from a fraction of the mass of the Sun to a million times as great.

There are several distinct types of nebula:

- *Emission nebulae*: clouds of very hot gas associated with regions of star formation and very young stars of age less than 10 million years. The clouds fluoresce,[†] illuminated by ultraviolet light from the massive young stars that have formed inside the clouds. These nebulae, which are the seat of very active star formation, are also called *H-II regions* for the large amount of ionized atomic hydrogen they contain (astronomers refer to normal, unionized atomic hydrogen as *H-I*, and to ionized hydrogen as *H-II*).
- *Reflection nebulae*: clouds of relatively cold gas and dust which would normally be dark, but whose dust grains reflect the light of nearby stars. These nebulae are often associated with regions of star formation and with moderately evolved stars, like those in the Pleiades cluster.
- *Dark nebulae*: similar to reflection nebulae, but seen against the light of more distant stars. Observation at infrared and radio wavelengths often reveal star formation inside them. A particular case is the *Bok globule*, an extremely dense, cold cloud of gas (among the coldest objects in the Universe, with temperatures of only a few tens of kelvin) which may eventually collapse and form a protostar.
- *Planetary nebulae*: bubbles, shells, or jets of hot gas fluorescing like emission nebulae, but this time excited by the ultraviolet emissions of a single, evolved star. Often spectacular in shape, they are produced by ejections occurring during the final evolutionary phases of stars of less than eight solar masses (Q. 15).
- *Supernova remnants*: clouds of stellar material expelled during supernovae explosions (Q. 16). This material collides at high speed (1% of the speed of light) with interstellar gas and makes it glow. The glow can be seen for centuries after the explosion, sometimes in the visible, but more often only at radio wavelengths or in x-rays.

These several types of true nebula, clouds of gas and dust located within our galaxy, should not be confused with what used to be called "nebulae" until the early twentieth century. The latter were actually galaxies outside our own, in which individual stars could not be resolved at the time.

[†] Fluorescence occurs when atoms are excited by ultraviolet light and re-emit the absorbed energy as visible light.

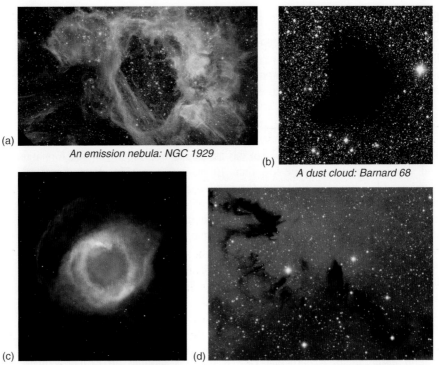

(a)
An emission nebula: NGC 1929

(b)
A dust cloud: Barnard 68

(c)
A planetary nebula: the Helix

(d)
A reflection nebula: NGC 6559

Different types of nebula: (a) NGC 1929 is an emission nebula in the Large Magellanic Cloud, 150 000 LY away. Credit: Gemini Obs.

(b) Barnard 68, at a distance of 410 LY, is a fine example of a Bok globule. The cloud looks dark in visible light, but is bright at mid- and far-infrared wavelengths. Interstellar dust hides the stars inside this dark cloud. Credit: ESO.

(c) The Helix, one of the closest and most spectacular planetary nebulae (680 LY). In this example, there are actually several envelopes, corresponding to successive ejections, associated with the final spasms of the central star which is about to be transformed into a white dwarf. Credit: O'Dell *et al.*/NASA/ESA.

(d) NGC 6559, the cosmic "dragon," nicely illustrates the environment of reflection nebulae which, rather than emitting light, reflect that of neighboring stars or hide more distant ones, so appear as shadow figures. Credit: Gemini Obs.

158 How empty is space?

Interstellar and intergalactic space, the space between stars and between galaxies, is not absolutely empty. It contains gas, mainly hydrogen, at extremely low densities. The material is as sparse as a few hydrogen atoms per cubic centimeter between galaxies and a few tens of thousands of atoms per cubic centimeter between stars. For comparison, the air we breathe contains about 10^{19} molecules per cubic centimeter. The temperature of this interstellar or intergalactic gas varies greatly, depending on the sources of heat nearby. Far from any star or galaxy, it is just a few kelvins (Q. 133). In the vicinity of

X-ray mosaic image of a region 400 x 900 LY across, showing the central region of the Milky Way (photo obtained with the Chandra X-ray Observatory). Hundreds of white dwarfs, neutron stars, and black holes swimming in a fog of hot interstellar gas at temperatures of several million K can be seen. Credit: NASA/Univ. Mass/D. Wang *et al.*

hot stars, it is about 10 000 K. In clusters of galaxies there are colossal amounts of extremely hot gas at about 100 million K which produce abundant x-ray emissions.

Interstellar space also contains dust. Although the mass of this dust is tiny (one hundred times less than the mass of the gas), these microscopic grains play an important role as storehouses for many of the heavy elements of the interstellar medium, such as iron, magnesium, and calcium. These dust grains absorb and scatter the light from even more distant stars and galaxies, reducing their apparent brightnesses.

Intergalactic and interstellar gas and dust can be detected during spectroscopic observations of distant bright sources such as hot stars and quasars. Superimposed on the characteristic spectra of the observed objects appear narrow spectral absorption lines that originate from hydrogen atoms and other ions and atoms located along the line of

Optical/radio composite image of the spiral galaxy M 51. In white, the Galaxy imaged in visible light: the light is produced by the stars, fluorescent gas, or reflection from dust. In blue, superimposed, the Galaxy imaged in radio waves at the 21 cm wavelength: this radio emission is produced by cold neutral hydrogen gas (a few tens of kelvin or less). The extent of the Galaxy at radio wavelengths is strikingly larger than is visible. Credit: NRAO/AUI.

sight. Since these spectral lines are subject to the cosmological redshift (Q. 141), such observations even permit the reconstruction of the chemical evolution of the Universe over time.

Interstellar and intergalactic space is also crisscrossed by very high-energy particles called cosmic rays. These particles are accelerated in processes taking place in the atmospheres of eruptive stars, in the magnetic fields of supernova envelopes, and in the nuclei of active galaxies. Some are re-accelerated in the magnetic fields of our galaxy and reach phenomenal energies (10^{20} eV). Cosmic rays contain many particles of antimatter, a telltale sign of the strange mechanisms that produce them (Q. 150).

"Empty space" is suspected also to contain dark matter and dark energy, but we know next to nothing about them (Q. 147, Q. 148).

159 How did the theory of relativity affect astronomy?

Albert Einstein
(1879–1955).

Albert Einstein did not really attempt to solve astronomical problems, but his work had an immense impact on cosmology in particular.

His theory of relativity consists of two great chapters, *special relativity* (1905) and *general relativity* (1915), and they cover quite different aspects of physics. These two theories have different fields of relevance and would probably have different names if they were not from the same author.

In his theory of special relativity,[†] Einstein sets two premises: (a) the speed of light is finite and constant, (b) the laws of physics are valid in all moving frames of reference.[‡] From these, he derived new laws of dynamics: length contraction, time dilation, and his famous equation of equivalence between mass and energy, $E = mc^2$.

The theory of general relativity is a geometrical theory of gravitation in which Einstein describes gravity, not as a force, but as the result of matter causing space-time to warp (Q. 160).

The theory of relativity does not contradict classical Newtonian mechanics, but encompasses it. Newton's laws are still valid at low velocities. However, for large masses and speeds that reach a significant fraction of the speed of light, the theory of relativity predicts phenomena unheard of in Newtonian physics, but which have been tested repeatedly and demonstrated to be real.

Huge masses (sometimes very dense ones) and extreme velocities are common in the Universe, making this the most fertile ground for relativity. Here are some examples of phenomena that could be understood only within the framework of relativity:

- The expansion of the Universe follows directly from the equations of general relativity (Q. 129). Just as a pencil standing on its point cannot remain in equilibrium, the Universe cannot remain static: it must either contract or expand. Observations indicate that it is expanding. Furthermore, it was discovered late in the 1990s that this expansion is accelerating due to the presence of dark energy (Q. 148). Although

---

[†] Special relativity is limited to uniform relative motion and does not take into account accelerated motion (this is covered by general relativity).

[‡] In physics, a frame of reference is a system of coordinates used to locate objects in space and time.

Gravitational lensing produced by the intervening massive cluster of galaxies Abell 2218. The distorted, arclet-shaped images are of galaxies that are much more distant than the Galaxy cluster producing the gravitational field that modifies their light path. Credit: NASA.

the nature of dark energy is unknown, it may correspond to one of the terms in the general relativity equations that describe the Universe.

- Black holes represent extreme cases of the warping of space-time by high concentrations of mass, warping it to the point where nothing, not even light, can escape (Q. 145).
- When combined with quantum physics, Einstein's theory explains some of the strange properties of very dense cosmic objects such as neutron stars and white dwarfs.
- The slowing of clocks in strong gravitational fields (gravitational redshift) accounts for the unusual behavior of light and radio waves produced by very massive objects.
- The distorted images of very distant galaxies viewed through some very massive clusters of galaxies are a spectacular manifestation of the gravitational lensing predicted by Einstein (see image). This same effect is responsible for the multiple images of quasars produced when their light beams graze past intervening galaxies – resulting in true gravitational mirages. This effect is called gravitational lensing. The amplification of brightness from this effect creates a natural "telescope" that can also be used to detect exoplanets (Q. 182).

160 What is meant by "four-dimensional space?"

Our everyday world has three dimensions: front-back, left-right, up-down. Hence, traditionally, the position of an object in space is described using a coordinate system with three axes perpendicular to each other.[†] The position of an object in space is thus completely described by three numbers, its x, y, and z coordinates. If the object also

---

[†] This system, which was proposed by French mathematician and philosopher René Descartes, is called the *Cartesian frame of reference*.

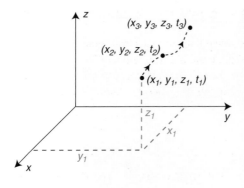

moves, a fourth number, time, t, is added to describe the movement of the object over time, and if an event is to be described, the same system of four numbers is used: three spatial coordinates to indicate where the event took place, plus a coordinate for time to define the moment when it took place. We can thus extend the geometrical concept of position in space by adding time as the "fourth dimension" because, just as for the three spatial dimensions, it is used to describe events and objects in motion.

So although the "fourth dimension" is a favorite theme of science fiction writers, there is really nothing mysterious about it. The issue is simply one of semantics. Albert Einstein himself found it ironic that "the non-mathematician is seized by a mysterious shuddering when he hears of 'four-dimensional' things, by a feeling not unlike that awakened by thoughts of the occult. And yet there is no more common-place statement than that the world in which we live is a four-dimensional space-time continuum."

What exactly is meant by the term "a continuum of space-time?" The four dimensions – three of space plus one of time – are independent of each other, are they not? The answer to that is no. The three spatial dimensions are certainly independent of each other, but when we add time, there is a snag. The problem is this: the only way we can know about an event is to "see" it, but as the speed of light, although very great, is not infinite, an observer positioned close to an event will notice it before a more distant viewer, i.e. light will take a longer time to reach and inform the more distant one. If there are several observers along the line of sight, each one will assign a different time to the same event, depending upon each person's location. Yet the event itself took place at a single instant. Therefore, it is not possible to know, in an absolute sense, when an event took place. Hence, time is not an independent dimension: it depends on space. This is the rationale behind the concept of space-time.

The basis of Einstein's relativity theory is this new understanding that there are no absolute frames of reference; that all frames of reference are relative and particular to each observer. In classical physics, time and space are treated as independent, and gravity, the force of attraction between objects with mass, is independent of space. In the theory of relativity, time loses its independence, and measurements of length and duration depend on the velocities of both the observer and the measured object. Furthermore, gravity warps space-time in that the presence of a mass deforms space and alters time.

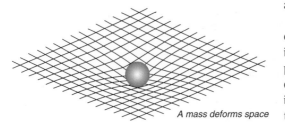

A mass deforms space

As a dimension, time remains very special, however. As far as we are concerned, it can move in only one direction, from past to future. Also, there can be only one dimension of time. It might be possible to imagine a type of space-time with only two – or maybe more than three – *space* dimensions, but there cannot be a space-time with zero – or more than one – *time* dimension.

Finally, according to string theory, space-time does not have four dimensions, it has ten or eleven of them: nine or ten spatial dimensions and one time dimension. Where are all those extra space dimensions to be found? According to the theory, they are compressed to such an extent that they are invisible in everyday life or in experiments we might devise (Q. 139).

161 Can anything go faster than the speed of light?

According to the theory of relativity, no material object can move faster than the speed of light as it travels through a vacuum, since that would require an infinite amount of energy and the mass of the object would also become infinite.

There are several apparent exceptions to this rule which are only illusions. If you point a powerful laser at the Moon, for example, and, with a flick of the wrist, move the bright spot across its face in a hundredth of a second, the spot will have traveled the distance of the Moon's diameter (3400 km) in 1/100 s, i.e. at 340 000 km/s, which is faster than light. This sort of situation actually occurs in the case of certain quickly rotating pulsars which project beams of radiation onto dust clouds – like lighthouse beams sweeping across the sea. The key is that these spots have no physical existence and carry no information with them. No *real* object is moving faster than light there.

Another case of a faster-than-light illusion occurs when objects located close to the line of sight move almost at the speed of light (see figure).

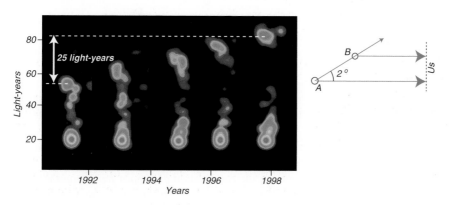

Sequence of radio images of clouds of high-energy particles (in blue) ejected from the quasar 3C279 (in red), recorded between 1991 and 1998. The clouds appear to move away from the quasar at super-luminal speed (25 LY over a period of 7 years). This is an illusion caused by the high velocity of the clouds (0.997 the speed of light) and in a direction only 2° away from the line of sight. A cloud appearing at position A at the start of the ejection is at B later on. However, since it is then also closer to Earth, we receive its light earlier, which leads us to underestimate the time involved while moving from A to B, hence to overestimate its speed. Credit: Anne Wherle *et al.*/NRAO/AUI.

But are there any examples of things that really go faster than light?

One way to do that is to make the light slow down, as happens when light travels through a medium other than a vacuum. High-energy cosmic ray particles can do this. They travel in space at close to the speed of light in a vacuum, but if they shoot through the atmosphere and plow into a tank of water, they continue at their original speed because they do not interact with the water. On the other hand the velocity of light in water is reduced to 75% of its value in a vacuum because it does react with water and other transparent media. The "super-luminal" speed of the cosmic rays is thus real, here, but the particles do not move faster than light moves in a *vacuum*.

Another example of objects that really move faster than light is the entire population of galaxies beyond our visible Universe. Galaxies recede from us at a velocity that is proportional to their distance (Q. 129). At 13.7 billion light-years, our cosmic horizon, they reach the speed of light. Beyond, they go faster than light, and become invisible to us. But their motion is not a motion in the ordinary sense. They are not moving through space; it is space that is carrying them along as it expands. So here, again, this is not in contradiction with the "thou shalt not go faster than light" commandment.

Not esoteric enough for you? Then think about the hypothetical massless particle, the *tachyon* (from the Greek for "swift"), which is posited to move faster than light – better yet, that cannot slow down to less than the speed of light. Although mathematically compatible with the theory of relativity, this particle has never been observed.

162 Why does everything in the Universe rotate?

The Earth rotates. The Moon rotates and orbits around Earth. All the objects of our solar system rotate and orbit around the Sun. Our sun rotates. Our galaxy rotates. The other galaxies rotate.

These rotational movements of celestial bodies and systems stem from the gravitational collapse of the primordial gas and dust clouds from which they formed.

If a cloud started out perfectly homogeneous in density, perfectly spherical in shape, and totally immobile, gravity would simply cause it to contract radially towards its center. In practice, however, clouds of dust and gas in space have random shapes, are not homogeneous, have chaotic internal motions, and may even be under the gravitational influence of nearby celestial bodies. As the clouds contract, asymmetries and internal collisions will impart organized motion in a global rotation, just as asymmetries and tiny residual currents in a tubful of water cause the water to rotate in one direction or another when the plug is pulled, rather than go straight down the drain.[†]

As the cloud continues to contract, the *conservation of angular momentum* accelerates the rotation, just as the skater who brings her arms in close will send herself into a dazzling spin.

There are certain systems that do not rotate or do so only slightly. This is the case with nebulae, where turbulent and chaotic motions dominate and large-scale rotation is rarely seen. Globular clusters also exhibit little or no overall rotation. The thousands of stars that make them up are, indeed, in orbit around their common centers of

---

[†] The belief that it rotates preferentially in one direction due to the Coriolis force is erroneous. The effect is much too small to be noticeable in a tub or sink.

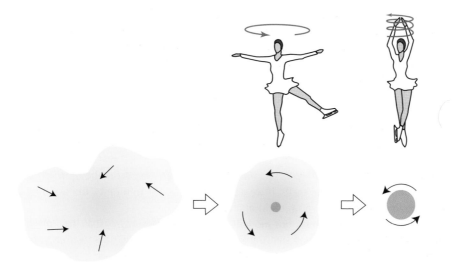

mass, but the stars' individual orbits have random orientations and their directions of revolution are also random. The stars in the cluster do not collide frequently so there is no mechanism to organize motion in global cluster rotation. Giant elliptical galaxies are spheroidal systems which behave like globular clusters and their individual stars have orbital motions in all orientations and directions, while the system displays little or no global rotation.

The Universe as a whole does not rotate. Any primordial spin it may have had at the beginning was completely erased during its inflationary period when, to use the analogy of the skater again, it stretched its arms wide open, to infinity, bringing its nascent spin to a halt.

163 Why is the night sky dark?

The sky is black at night. If you stop to think about it, that is really rather surprising. Since the Universe is huge, even infinite, and filled with billions of stars and galaxies, when we gaze out in any one direction our line of sight ought inevitably to encounter the bright surface of some celestial body, no matter which way we look. Think of standing in the middle of a good-sized forest: look in any direction and your gaze meets a tree trunk. The forest seems limitless and you can see no way out. The sky, too, is limitless in appearance, and we know that there are so many stars out there that the night sky ought to be uniformly brilliant. Why isn't it?

This apparently simple question, called "Olbers's paradox" after the German astronomer who formulated it in 1826, had astronomers perplexed for more than a century. Olbers made one assumption, that matter is uniformly distributed in the Universe, then reasoned as follows. If we consider all the stars that are equally distant from Earth inside of what we might call a "shell" of space, they send a certain total amount of light energy to Earth or, more precisely, we receive a given number of

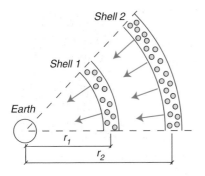

photons per second per square meter from them. If we consider a second, more distant shell, equally thick, the volume occupied by the stars in it is larger (by the square of the ratio of their radii, $(r_2/r_1)^2$) and the number of stars is therefore greater. So even though each star may appear less bright because it is more distant (also by the square of the ratio of their radii – see Q. 6), there are more of them, and we actually receive exactly the same amount of energy from the large, distant shell as from the smaller, closer one. Since the energy received per second and per square meter is the sum of the contributions of each of the shells, if we add more shells out to infinity, our sky ought to be extraordinarily bright, both night and day.

What Olbers could not know, but modern astronomy has shown, is that the Universe is not limited to our galaxy, and that the finite velocity of light prevents us from seeing the Universe in its entirety. Although it is infinite in size, we cannot see anything more distant than our cosmological horizon (Q. 142), which is 13.7 billion LY away. Inside this visible Universe, there are too few stars and galaxies to fill the sky.

Olbers was also making the implicit assumption that stars never die – but they do. Several generations of them have already come and gone since the Big Bang. So at any given time, there are simply not enough stars "alive" to cover our sky.

The expansion of the Universe also changes the game. As light from stars and galaxies propagates through an expanding space, the light expands, too: it "lengthens," its frequency diminishes, and its energy (which varies as its spectral shift is raised to the fourth power) drops dramatically with distance. The sky background due to starlight is thus very cold, hence of low luminosity – 10^{12} dimmer than the surface luminosity of an average star.

Therefore, it is dark at night because stars do not live for ever and because the Universe is expanding. Isn't it surprising how such a simple question cannot be answered without a profound understanding of the very nature of the Universe?

Nevertheless, even on a moonless night with no city nearby, the sky is not really black. The light from the stars may be dim, but it is not negligible. Try the following yourself: some dark night, stand in front of a white wall with the Milky Way behind you. You will see your shadow projected on the wall – albeit faintly – by the light of the stars.

164 What is the anthropic principle?

The world appears to be made for us. In many mythologies and religions, the Earth is at the center of the Universe, created for our express benefit. Although the Copernican revolution destroyed this simplistic view, certain discoveries of contemporary astronomy have actually caused this centuries-old centric notion to regain popularity.

It is true that the Universe appears to be fine-tuned to favor the emergence of life. If the fundamental physical constants and the initial conditions of the Big Bang had been slightly different, the Universe as we know it would never have been (Qs. 149 and 150). If the expansion rate had been different, life could not have arisen (see figure). If water molecules and atoms of carbon had been different, life could not have evolved (Qs. 174 and 175).

Are these just coincidences? Theoreticians in various disciplines (high-energy physics, astrophysics, mathematics) have attempted to explain the observed characteristics of the Universe through a concept called the "anthropic principle." This principle, proposed in its modern version in 1974 by the Australian theoretician Brandon Carter, exists in two main versions, *weak* and *strong*. The weak version of the anthropic principle proposes that the Universe is governed by finely adjusted physical laws that allow for life and consciousness. The American physicist Freeman Dyson summarizes it this way: "*The Universe knew we were coming!*" The strong version affirms that the fine-tuning does not arise from sheer chance, but is the product of some cosmic finality. This version is popular among some, especially the religiously or mystically inclined, because it appears to support the idea of a Universe created by a "master architect."

Brandon Carter.

If the weak version is a simple acknowledgment of fact – the Universe must be compatible with the emergence and evolution of life because we do exist – the strong version is an unscientific lazy escape, invoking a supernatural explanation for a challenging problem.

First, is it really true that, for life to exist, our finely tuned set of physical laws and environmental conditions is necessary? For example, could the presence of liquid water and carbon-based organic chemistry be dispensed with? Maybe so. As Fred Hoyle posited in his fine science-fiction novel *The Black Cloud*, there might be a form of intelligent life out there composed of a giant interstellar cloud – or a form that no one has yet dreamed of. So far, we have barely explored one of the 10^{22} solar systems calculated to exist in the observable Universe (roughly 10^{11} galaxies, each with 10^{11} stars per galaxy). What surprises still await us? Maybe the requirements for life are a lot less strict than we imagine. Maybe the Universe is not so fine-tuned after all.

If the Universe expands too slowly, it soon collapses, leaving too little time for stars to produce the basic elements of life (red/orange region). If the expansion is too rapid, gravity cannot condense matter enough to permit the formation of stars and galaxies (blue region). The zone favorable to the existence of life is in between (white region).

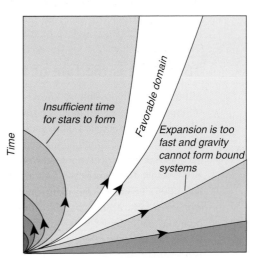

Size of the Universe

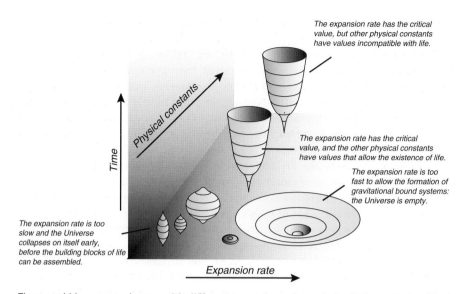

The expansion rate has the critical value, but other physical constants have values incompatible with life.

The expansion rate has the critical value, and the other physical constants have values that allow the existence of life.

The expansion rate is too fast to allow the formation of gravitational bound systems: the Universe is empty.

The expansion rate is too slow and the Universe collapses on itself early, before the building blocks of life can be assembled.

Time

Physical constants

Expansion rate

There could be many universes with different expansion rates and physical constants with different values, but only a few would permit the emergence of life. Credit: S. Hawking [25].

But if it is true that at least our portion of the Universe is finely tuned for life, then string theory (Q. 139) proposes an explanation that erases the statistical improbability of our uniqueness. In this theory, the cosmological constant varies from one part of the Universe to another (Q. 147). Our own portion of it would be a particular case favorable to life; in other portions, that would not necessarily be the case.

Pushing string theory even further brings us to the multiverse concept (Q. 138). Our universe would be merely one of the 10^{500} possible combinations of physical constants and expansion rates. While our own universe is favorable to the existence of life, others would not be.

165 What is the fate of the Universe?

Although it is impossible to know with any certainty what will happen in hundreds of trillions of years, we do have the tools to make a few predictions. Two assumptions are adopted: (a) the accelerated expansion that has recently been discovered will continue forever and (b) the fundamental constants will not change (Q. 149). Alternative theories such as that of a cyclic universe would make different predictions.

We know that when the Sun reaches the age of about 10 billion years (10^{10}) it will briefly become a red giant on its way to becoming a white dwarf (Q. 38). For a while, however, other stars will continue to shine in our galaxy and new stars will form here and there. In a few tens of billions of years, the Milky Way, the great Andromeda Galaxy, and the Magellanic Clouds will have become one single giant elliptical galaxy. Elsewhere in the Universe, the other galaxies with close neighbors also will have merged into super-galaxies. These giant cannibal galaxies will be bright, but they will also have receded so far from one another as the Universe expands, that an observer in any one of them would be unable to see other galaxies, even its closest neighbors.

The amount of hydrogen available in the Universe will diminish over time, consumed in successive generations of stars. The rate of star formation will slow and stop. Then, one by one, the last stars will cease to shine. The longest-lived stars will be those of very low mass, 10 times less than that of the Sun or even lower, because such bodies convert their hydrogen very slowly. After hundreds of billions, perhaps even trillions, of years, their hydrogen finally exhausted, these, too, will slowly cool and the last sources of light and heat will be extinguished.

By then the Universe will be 10^{16} years old, a million times its current age. The dominant forms of matter will be brown dwarfs, dead stars (white dwarfs, neutron stars), and black holes. At the age of about 10^{25} years, these strange objects making up the galaxies will either be captured by black holes or ejected into the intergalactic void. Galaxies will then only contain degenerate dwarfs and relatively low mass black holes, with the exception of the galactic cores that will each harbor a supermassive black hole. For any future hypothetical astronomers, the sky will be dark and empty.

Finally, even fundamental particles are unstable over long periods of time. Protons will decay after 10^{30} years, leading to the disintegration of the brown dwarfs, white dwarfs, and neutron stars. Only electrons, positrons, neutrinos, and a weak background radiation will be left. Black holes will still exist, but even they will not last for eternity: they lose Hawking radiation (Q. 146). Their masses will diminish, slowly at first, then faster and faster as their masses dwindle. The largest black holes, with masses of the order of that of a galaxy, will take 10^{100} years – an astounding amount of time – to evaporate. In this final era of evanescing black holes, the tranquility of the Universe will be shattered by rare bursts of light and high-energy radiation produced by the death throes of the last dark colossi.

And then ... nothing. That will be the end of it, the end of everything. Not a very cheerful story, is it? But then, no one will be around to deplore the sad ending. All life will have disappeared long before.

166 What major questions remain to be answered in astronomy?

The twentieth century is viewed by many as having been a true golden age of astronomy, and we now have the answers to most of the questions that have pre-occupied the human mind for millennia. The origin of the Earth, the nature of the Sun and of the Milky Way, the origin and evolution of the cosmos, are no longer so mysterious.

As we continue to unveil the Universe, however, we discover its stunning complexity and new enigmas appear. Here is a short list of big questions that remain to be answered.

• How much did collisions with comets and asteroids affect the evolution of life on Earth?
• Is there any manifestation of life elsewhere in the solar system?
• Is there intelligent life elsewhere in the Universe?

- Is our understanding of the Universe significantly limited by shortcomings in our current grasp of physics?
- What is the nature of dark matter?
- What is the nature and origin of dark energy?
- How can we prove the existence of black holes?
- Does the Universe have more than four dimensions?
- Is ours the only universe?
- Is our current view of the destiny of the Universe correct?
- How could the hypothesis of the multiple-universe be verified?

167 How can we hope to comprehend the *astronomical* numbers which astronomy confronts us with?

The ages, distances, sizes, and masses of stellar objects are, indeed, astronomically large. How can the mind grasp such enormous values? Our short lives of less than one hundred years and the sluggish speeds of our space vehicles leave us no hope of traveling further than one light-day, the approximate size of the solar system. This is a minute fraction of the observable Universe, which is 10^{36} (1 followed by 36 zeros) times larger.

Nevertheless, even the vast dimensions of the Universe are relatively easy to conceive. When we look up at the night sky, we see stars that are a few light-years away (e.g. Sirius or Vega), others that are hundreds of times more distant. When we gaze at the Milky Way, our eyes encompass a cosmic landscape that stretches over many thousands of light-years. Yet these vast expanses of space do not necessarily overwhelm us. Even out beyond our galaxy, the numbers required to describe the observable Universe, with its scale of billions of light-years and its age of almost 14 billion years, are generally still comprehensible. Million, billion, even trillion – these are numbers that appear frequently in today's media – their values can even be counted as modest in the context of the

Large numbers: the supergiant elliptical galaxy NGC 3311 (center of the image), 176 million LY away, contains more than one trillion stars. The Galaxy is surrounded by a suite of 16 000 globular clusters, each containing hundreds of thousands of stars over 10 billion years old. Credit: Wehner *et al.*/Gemini Obs.

budgets of certain large countries. The Earth has a population of nearly 7 billion. Our Sun, Earth, and Moon, at 4.5 billion years of age, are relative old-timers; in certain regions of our planet you can walk on rocks that formed 3 billion years ago.

However, the mega-universe proposed by string theory blows away our familiar concepts: the numbers that it generates are staggering. Our universe is only a minute fleeting bubble in a foam of infinite and eternal universes. Blaise Pascal said it for us long ago: "*the silence of these infinite spaces is frightening*." Even for the seasoned astronomer, it is exceedingly difficult to grasp the concept of 10^{500} possible universes of which ours would be but one of the statistical realizations.

At the other end of the scale, in the ultra-miniature world, we find ourselves facing the smallest dimensions the physicist can describe. The Planck length, 10^{-33} cm, when compared to the size of an atom (10^{-8} cm), is as small as an atom compared to the size of a galaxy. The graviton, the messenger particle of the gravitational force, is proposed to be the size of the Planck length. Planck time, 10^{-43} s, is the smallest lapse of time allowed by quantum mechanics. The ratio between Planck time and 1/10 000th of a second is the same as between 1/10 000th of a second and the age of the Universe. As enlightening as these comparisons can be, they fail when one attempts to deal with numbers with exponents greater than 100.

Whether they fascinate or frighten, astronomically large numbers reveal a fundamental characteristic of nature: *diversity*. In our Universe, the properties of the fundamental forces and of elementary particles have combined to produce a cosmic landscape of incredible richness: gigantic masses of gas at 100 million K, clusters of thousands of galaxies, hellishly hot stars, brown dwarfs, icy asteroids, clouds of incredibly cold dust. There are also perhaps as many as 10^{22} planets in the observable Universe (Q. 164). This must give rise to many possible environments favorable to life in general and to intelligent life in particular. Some planets are surely biological dead-ends – or non-starters – but given even a small chance, life proves to be tough. On Earth, it has already managed to stay in existence for at least 3 billion years. There is no magic involved in that: the physics of large numbers allows nature to explore all kinds of possibilities. It is they that endow the Universe with its extraordinary variety.

168 Is there a difference between the cosmos and the Universe?

The terms *cosmos* and *universe* are often used interchangeably, but there is a subtle difference of meaning.

In ancient Greek, *kosmos* represented the idea of order, of organization, and also of decoration (hence, *cosmetic*). For philosophers, it meant the order of the world, and in late Greek, the inhabited world, and humankind. It acquired its current meaning relatively recently, from the title of a massive, celebrated work by the great German naturalist Alexander von Humbolt (1769–1859): *Cosmos: Essay on a physical description of the world.* Since then, *cosmo-* has generated a number of new terms: cosmography, cosmology, cosmogony, and, most recently, cosmobiology. When the Russians made their spectacular entry into space exploration in the 1950s, they gave the word the meaning of extraterrestrial space, calling their astronauts *cosmonauts*.

The vision of the world developed during the Middle Ages involves the concept of order, cosmos. The terrestrial world and human activities are relegated to the "sublunary" world that also encompasses atmospheric phenomena, meteors, and comets, while the celestial world, viewed as absolute perfection, floats above it. Credit: Les très riches heures du duc de Berry, Musé Condé, Chantilly.

The word "universe" comes from the Latin, *universum*, "made one" (from *unus*, one, and *vertere*, turned, or changed to). The astronomical meaning of "universe" has changed over time, keeping pace with the evolution of cosmological concepts. At first limited to meaning "the surface of the Earth," the term expanded to encompass the solar system at the time of the Copernican revolution, and it now includes everything that exists.

Life in the Universe

169 What is life?

Are we alone? Does life exist anywhere besides Earth? Modern astronomy can help to answer these fundamental questions, but first we need to know what is meant by *life*.

The concept is hard to define. It used to be generally accepted that life is organized matter that exhibits seven crucial characteristics: growth, respiration (i.e. exchange of gases), nutrition, excretion, reproduction, reaction to external stimuli, and locomotion (the last is only partly true, since plants and many other organisms are not mobile). Matter in a crystal is organized, and one could argue that a crystal grows by nourishing itself with neighboring atoms, that it reproduces when it branches out, and that it reacts to external stimuli, contracting when subjected to an electrical charge, for example (as in the piezoelectric effect used in quartz watches). Nevertheless, a crystal does not breathe, does not excrete, and does not move. Then what about fire? A fire nourishes itself with combustible material, absorbs oxygen (breathes), grows, moves, can ignite additional fires in new locations (reproduces), excretes heat, and reacts to external stimulations such as wind or a fire extinguisher ...

But if it can be argued that fire meets the seven basic criteria for life, living creatures fulfill one additional condition: they can mutate. This allows an organism to adapt to new environments and thus to *evolve*. Fire cannot evolve into a different form of oxidation, such as rusting, but we know from the fossil records that life transforms itself all the time, resulting in the extraordinary variety of forms around us: from bacteria to watermelons to oysters to elephants.

So the old definition of life, which was both too restrictive and incomplete, has now given way to a new one that could also eventually apply to life elsewhere in the Universe: what fundamentally differentiates life from inert matter is its capacity to *grow*, *react* to external stimuli, to *reproduce*, and to *evolve*.

On Earth, the powerful and versatile DNA molecule enables all these functions to be performed: reproduction by making copies of itself, evolution by mutating, and growth and reaction to external stimulation by programming its host to grow and react.

Amorphous

Organized, grows, reacts

Organized, grows, reproduces, reacts, evolves

170 How did life begin on Earth?

The oldest fossils known date from about 3.5 billion years ago. Simple bacterial forms that lived in shallow water, they closely resemble certain bacterial strains alive today [39]. Since the fossil bacteria must necessarily have evolved from more primitive life forms, it is generally agreed that life must have first appeared on Earth around 3.8 or 4.0 billion years ago, which is surprisingly soon after our planet's formation 4.5 billion years ago. Yet it cannot have appeared much earlier than 4 billion years ago because Earth was subject to heavy bombardment during the first half billion years of its existence (Q. 33) and was too hot for liquid water to exist. The earliest life was a simple cell, and how that first cell came into being is not yet entirely understood, but we can posit some likely steps.

Looking first at the chemical composition of living matter, we find that it is essentially made of carbon, nitrogen, oxygen, and hydrogen, which happen to be the four most abundant elements in the Universe (discounting helium, which is an inert gas and does not combine). These four furnish fully 96% of life's constituents, the remaining 4% being composed of phosphorus, sulfur, and traces of other elements. The high concentration of oxygen and hydrogen in living matter is easily explained by the water it incorporates, and by the high reactivity of oxygen. But although carbon and nitrogen are found in stars in significant amounts, they are not at all abundant on Earth, which is mostly made of silica and iron. Why, then, would the recipe for life call for carbon and nitrogen? Is there something special about their chemical behavior?

The chemical reactivity of an element depends on the number of electrons in its valence (outer) shell, with eight electrons being the upper limit. Carbon, with four electrons in its valence shell, can form up to four bonds with other atoms, each bond consisting of one electron from the carbon atom and one from the atom with which it

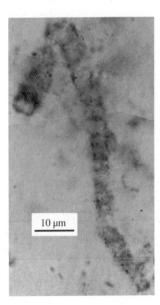

One of the oldest fossils known, a 3.5 billion year old bacterium from a rock formation in Australia.

10 μm

Top, an imaginary molecule composed of atoms, represented by Xs, which can only bind to two other atoms. Such atoms can only form linear molecules. Carbon, which can bind to four other atoms, can form complex molecules such as glucose, shown at bottom.

$$Y - X - X - X - X - Z$$

$$\begin{array}{c} O \\ \diagdown \\ C \\ \diagup \\ H \end{array} \begin{array}{c} OH \\ | \\ - C \\ | \\ H \end{array} \begin{array}{c} H \\ | \\ - C \\ | \\ OH \end{array} \begin{array}{c} OH \\ | \\ - C \\ | \\ H \end{array} \begin{array}{c} OH \\ | \\ - C - CH_2OH \\ | \\ H \end{array}$$

bonds. Carbon can bond with hydrogen to form hydrocarbons, but can also bond with itself and with many other chemical elements. This allows an extraordinary variety of complex molecules to form and to be used in the basic processes of life: building cell structures, storing energy, encoding the information necessary for reproduction, etc.

Nitrogen, too, can share more than one electron with a carbon atom to form strong but breakable bonds to build a large number of diverse, stable molecules. Moreover, nitrogen gas, which is highly volatile, can easily cycle between organisms and their environment.

So it is because of their remarkable chemical properties and not by mere chance that hydrogen, carbon, oxygen, and nitrogen became the elements around which life was built.

We must then ask, how did these four elements combine to form the many different organic (i.e. carbon-based) compounds of which living things are made: the essential amino acids, the DNA, the exquisitely complex proteins?

Some have proposed that large organic molecules – or even life itself in the form of bacterial spores – came to Earth from space, fully formed. The idea is not completely far-fetched, as the young Earth was heavily bombarded by meteoroids and comets which do sometimes contain organic compounds (Q. 173). But aside from the fact that such a view merely pushes the question of origins back one step (the molecules did have to evolve somewhere), it is most probable that the vital molecules were synthesized quite naturally on Earth. The four basic elements of carbon, oxygen, nitrogen, and hydrogen were, after all, the main components of Earth's primitive atmosphere – in the form of carbon dioxide (CO_2), nitrogen gas, water vapor (H_2O), and a small amount of free hydrogen. And, as a famous experiment by Stanley Miller and Harold Urey has shown, many organic compounds – including amino acids, the basic components of proteins – are relatively easy to synthesize from such an atmosphere. Even if we have not yet managed to synthesize the proteins, the lipids, or the nucleic acids themselves, that early experiment would seem to have pointed the way down a promising path.

Nevertheless, for life to get started, it would never be enough simply to produce batches of elementary organic molecules; they would have to encounter each other and interact to form the more complex molecules of life. A liquid is a perfect medium for such interactions, provided that it is a good solvent so that the molecules diffuse well and do not precipitate out. And water, thanks to the polar nature of its molecule and its abundance on Earth, is ideal for that role (Q. 174). So it is generally assumed that life on Earth arose in water, which is an essential component of all known living cells and which, indeed, constitutes 70 to 95% of a cell's makeup.

Finally, a physical boundary between the living and non-living matter is needed, between the components of the cell and the hostile external environment. That role is

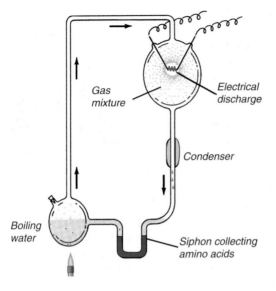

Gas mixture

Electrical discharge

Condenser

Boiling water

Siphon collecting amino acids

Schematics of the Stanley Miller and Harold Urey experiment which produces amino acids after a few days of continuous operation. In the original experiment, dating from 1952, the gases used were ammonia, methane, and hydrogen, then thought to be the main components of Earth's primitive atmosphere. The experiment has since been redone using carbon dioxide, nitrogen, and hydrogen and produces even better results.

played by a cell's protective membrane. Exactly how the first membrane formed around the nascent cell is still a mystery, but it is probable that the very first organisms did not fabricate their membranes themselves, but instead developed inside the naturally occurring envelopes, such as fatty bubbles, that tend to form in watery mediums.

Once life had become established, it remained essentially unicellular for a very long time – over 3 billion years. The first evidence for multicellular organisms is the enigmatic Ediacaran fauna that emerged in the Precambrian era about 635 to 540 million years ago. During the Cambrian era, from about 540 to 490 million years ago, more complex multicelled organisms developed. And soon after, approximately 530 million years ago, a veritable "explosion" of life forms with different body plans took place over a very short, 10 million-year span of time, producing the ancestors of most of the main animal lineages (the phyla) that exist today.

171 Does life violate the second law of thermodynamics?

Remember the first law of thermodynamics: the total energy of a closed system is always conserved. Although this law was originally developed for thermal systems, the principle of the conservation of energy applies to all forms of energy: mechanical (work), electrical, chemical, etc. The total energy of an isolated system – in this case, of the Universe – remains constant no matter what. One kind of energy can be transformed into another, but no energy is ever created or destroyed.

The second law, on the other hand, states that every such transformation from one form of energy into another is accompanied by a loss of *usable* energy. An electrical motor transforms electrical energy into mechanical energy, but its efficiency is not perfect and a small amount of electrical energy is lost in heat which is not usable. This loss of usable energy is called *entropy*. The concept was defined statistically by the Austrian physicist Ludwig Boltzmann as a measure of the state of disorder of a system.

The greater the disorder, the less usable energy there remains, and disorder increases inexorably with time. The entropy of the Universe has thus been increasing ever since the Big Bang, but although that is true of the Universe as a whole, nothing excludes the creation of order *locally*. Naturally, the local increase of order would be at the expense of a greater disorder outside it. This is what occurs where the phenomenon of life is concerned.

Living organisms are made of matter that has become organized. If the contents of a box of jigsaw puzzle pieces is thrown into the air, there is next to no chance that the pieces would land on the ground as a completed image. To put the pieces together properly, one would have to think, move them around, and interlock them, all

Ludwig Boltzmann (1844–1906).

of which uses energy. Similarly, life requires the selection of elementary molecules, their transformation into biomolecules, and their organization into cells. For complex organisms, the cells must then be organized into tissues, organs and, finally, into systems (circulatory, respiratory, digestive, nervous, etc.), then placed under the control of the brain. All of this consumes energy, and even more energy is needed to keep the organism alive, energy that is taken from the environment in the form of food and solar radiation [40].

So although people of a mystical bent may prefer to think that life is mysterious to the point of falling outside the realm of physical laws, it just isn't so. The global entropy of the Universe increases when an organism is created, it increases throughout the organism's life, and increases still further when it dies and its biomolecules decompose. Its internal order becomes disorder. The second law of thermodynamics is implacable.

172 Could intelligent life reverse the fate of the Universe?

The Universe is destined to expand forever, doomed to become a realm of extreme cold and extreme chaos, devoid of galaxies, of stars, of energy, where life is no longer possible (Q. 165).

Faced with such a woeful destiny, could some form of intelligent life possibly take over before it is too late, inject order into disorder, and recreate a Universe where life can exist?

Alas, no. As explained in the preceding question, organization within the Universe can only be had at the expense of greater disorder elsewhere inside it. One could imagine humanity creating a replacement sun when our own Sun dies, for instance, but this would only delay the inevitable. The entropy of the Universe cannot be reversed. The ultimate end is without appeal. You might find some comfort in reading Isaac Asimov's classic short story, *The Last Question* [2].

173 Could life on Earth have originated in outer space?

Nobody really believes that little green men came from a far-off planet to seed the Earth with life, but the idea that life could propagate from star to star, planet to planet, is not necessarily pure science fiction, either. The idea was proposed independently in the 1870s by two well-respected scientists, Hermann von Helmholtz, a German, and the Englishman William Thomson (Lord Kelvin). At that time, Louis Pasteur had just disproved the theory of spontaneous generation and Charles Darwin had just published his theory of evolution. Helmholtz speculated that if life could not arise spontaneously on Earth, it must have come from outer space. Lord Kelvin proceeded to do thermal calculations to estimate how long it would have taken the young Earth to cool down, and found that the planet was not old enough to allow for the long, slow evolution of life that Darwin's theory required (Q. 81). He therefore also proposed that living cells could have seeded on the Earth after traveling through space, perhaps inside meteoroids.

This idea, called "cosmic panspermia" (from the Greek *panspermia*, mix of seeds), remained pure speculation until Svante Arrhenius, a Swedish Nobel prizewinner in chemistry, showed in 1908 that some plant spores remained viable after exposure to an interstellar-like environment (high vacuum and temperature of −200 °C). Recent experiments have shown that certain types of bacteria can also survive strong doses of ultraviolet and cosmic rays such as are found in space, especially if tucked away inside a meteoroid. Several biologists even claim to have reanimated bacteria trapped for millions of years inside amber, ancient ice in the Antarctic [6], or in salt crystals [42].

The idea of panspermia was dusted off in the 1970s by the famous British astronomer Fred Hoyle (Q. 131), who argued that if bacteria can survive the environment of space, that must be where they originated! He contended that the clouds of interstellar gas and dust are repositories of dried-up bacteria, which can hitch a ride on passing comets and make their way to Earth. Panspermia would thus explain how life got its start on Earth so soon after the planet's formation. Advocates of panspermia argue that this mechanism is still at play today, interfering with the evolution of life on Earth, and that some of the great pandemics, such as the Black Plague, could be due to bacteria or viruses raining down from space.

Still, although panspermia is an attractive theory and hard to refute, there is no unequivocal, tangible proof to support it.

174 Why is water so important for life?

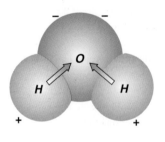

As we all know, water is composed of one atom of oxygen bound to two atoms of hydrogen (H_2O). When atoms come together to make a molecule, the attraction that keeps them together is electrical (what is called *valence* in chemistry), and the result of such a union is usually electrically neutral. But the water molecule is different in that it is not neutral *locally*. The lone electron of its hydrogen atom is attracted by the atom of oxygen, leaving its proton "naked" on the other side. This creates a strongly positive pole. And since the oxygen atom has now become overendowed with electrons, its "back" forms a slightly negative pole. The water molecule is thus dipolar and tends to be attracted to neighboring water molecules, with the naked hydrogen side of one molecule cozying up to the well-endowed side of the oxygen atom in another molecule, a particularity known as "hydrogen bonding."

The polarity of the water molecule gives it a property that is essential to life: it is an excellent solvent for substances that are not electrically neutral, especially for those that dissociate into ions, such as salts, and for polar molecules such as sugars and proteins. The water molecules are electrically attracted to the salt ions or to the polar regions of large molecules, surround them, and carry them around in their aimless Brownian motion, keeping them from precipitating. We say that such salts and molecules are "dissolved." Water can then transport them to their points of utilization (as in the case of blood or sap) and allow them to interact with other molecules to form complex molecules. Other solvents, such as ammonia and methyl alcohol, exist, but they are much less available in nature and are about twice as inefficient. Water is the *solvent of life*.

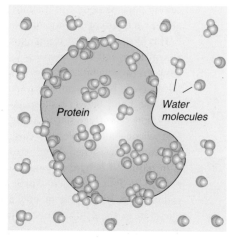

The polarity of the water molecule also explains why water expands when it freezes. In its solid state (ice), the lower temperature means less molecular movement: the molecules settle into a rigid crystalline structure and are glued into place by their hydrogen bonds. The combination of attractions and repulsions between molecules causes them to remain a certain minimum distance from their neighbors, creating small empty spaces, tiny holes, in the crystal lattice. In the liquid state, however, where kinetic (thermal) energy is higher, the molecules are free to orient themselves. Here, hydrogen bonding pulls them snugly together. And since these bonds are constantly breaking and reforming, the molecules end up, on average, closer together than when they are inside ice crystals: liquid water occupies less volume than ice: ice is lighter than liquid water.

This situation does not occur in most common substances because their molecules are not usually dipolar. Electrical forces do not keep the molecules in most solids apart, so they crowd together, making their solid phases denser than their liquid phases.

It is thanks to this oddity of physics that our oceans at high latitudes and many lakes remain liquid in winter. If ice were denser than liquid water, it would form at the surface in contact with the cold air, then would immediately sink to the bottom, where it would tend to build up rather than melt. Luckily for us, ice does float, stays at the surface, and insulates the liquid water beneath it, allowing aquatic life to survive.

Water has two other characteristics that make it advantageous to life: it stays liquid over a relatively large temperature range, 100 °C (versus 45 °C for ammonia), and its specific heat is high, making it relatively insensitive to changes in external temperature. This makes it a very favorable environment for – and component of – living things.

175 Could life evolve based on a chemical element other than carbon?

Life as we know it on Earth is based on carbon's ability to form complex molecules thanks to its four valence "arms" (Q. 170). Silicon, which is just below carbon in the periodic table of elements, has also a valence of four, and can form chains and rings as carbon does. Since silicon is almost as abundant as carbon in the Universe, might it not also serve as a base for life in some extraterrestrial venue?

This idea, beloved of science fiction writers, does not stand up to close examination. Silicon turns out to be far less favorable to life than carbon because bonding forces between two silicon atoms are only half as strong as those between carbon atoms, so a silicon chain is easily broken. Moreover, the bond between two silicon atoms is weaker than the bond between silicon and hydrogen, meaning that silicon/hydrogen chains with more than three silicon atoms are extremely unstable. That leaves next to no hope that silicon-based molecules could form in the extraordinary variety of rings, chains, and complex branched shapes that dominate organic chemistry.

In nature, it is hard to find even the simple molecule composed of a silicon atom bonded to four hydrogen atoms, *silane* (SH_4), the equivalent of methane (CH_4) for carbon. It is not found in meteorites, nor in the atmosphere of the other planets, nor even in the Earth's crust where silicon abounds. Silane has only been detected in the interstellar medium. Methane, on the other hand, and a host of other organic compounds have been found in the atmospheres of the other planets, in meteorites and, of course, on Earth.

176 What are extremophiles?

Extremophiles, as their name implies, are organisms that "like" extreme conditions, or even need them in order to survive. They are bacteria for the most part, but molluscs as big as dinner plates, crabs, and certain meter-long tube worms have also been discovered in the extreme environments found around the black smokers of abyssal waters.

Anaerobic bacteria that live without oxygen and for which oxygen is even a poison (which used to be true of all life, back in its earliest days) have long been known to exist. However, bacteria have also recently been found thriving under amazingly hostile conditions: very high temperatures ($+160\,°C$) or very low ones (ice), in strongly acidic or salty environments (salt lakes, salt crystals), under extreme pressure (3500 m deep), in the total absence of light, or during vastly extended periods of dryness. Amazingly, these organisms do not merely tolerate their lot – they do best in their punishing habitats and, in some cases, require the extreme conditions in order to reproduce.

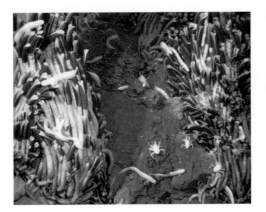

Fauna near a "black smoker" at a depth of 2600 m, where darkness is total. These creatures rely on energy supplied by the bacterial decomposition of sulfide into sulfur. Credit: Ifremer.

Bacteria have even been found in the hot, pressurized water of nuclear reactors! Some, like *Deinococcus radiodurans*, can survive intense radiation damage by repairing its own DNA, even when the radiation has fragmented its DNA into more than 100 pieces.

Extremophiles are found everywhere on Earth, and some biologists think that life on this planet started with cells of this type. Finding life in extreme environments here is encouraging, because it increases the chances that it could arise on other planets where conditions are much rougher, or at least could persist when conditions become more difficult, as they did on Mars.

177 Given favorable conditions, will life inevitably appear?

"If we are alone in the Universe, it sure seems like an awful waste of space," Carl Sagan once said. And, truly, conditions in the Universe do seem to be finely balanced to give life a good chance to appear (Q. 164). The expansion rate is tuned such that it has allowed several generations of stars to form, to produce the chemical elements of which living organisms are made (Q. 21), and to give biological evolution the time to come up with some interesting results: us, for example. If the rate was slower, the Universe would fall back onto itself and die too soon. If it was faster, matter would be spread so thin that it could not assemble to form stars. The physical constants that govern the Universe also seem to be finely adjusted for life: the carbon atom and the water molecule with their favorable characteristics are available (Qs. 170 and 174). If the mass and charge of the proton were very different from what they are, either the nucleus of the carbon atom would be unstable and fly apart or the electrons would collapse onto the nucleus. So our universe does indeed appear to be favorable to life. Like the porridge in Goldilocks's little bowl: not too hot, not too salty . . . just right.

But with the Universe so propitious to life, how can we Earthlings dare to imagine that it exists only on our planet? We may not yet have tangible proof, but with the discovery of numerous organic molecules in interstellar space and the large number of extrasolar planets already found, with 100 billions stars in the Galaxy and with 100 billion galaxies in the visible Universe, it is hard to doubt that life of some sort exists elsewhere.

178 Where in the Universe would life have the best chance of appearing?

The apparent lack of life in the solar system aside from our own planet, combined with what we know about life on Earth (Q. 170), suggests that life can only appear in a place

- where enough heavy elements (carbon, silicon, iron, etc.) are present for a rocky planet like Earth to form and for biomolecules to develop;
- where water exists in a liquid state to serve as a solvent for biomolecules;
- where there is not too much damaging radiation, e.g. from nearby supernovae and bright stars;
- and where bombardment by comets and other bodies is not too intense or frequent.

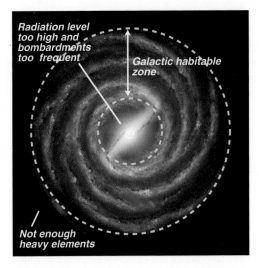

Radiation level too high and bombardments too frequent

Galactic habitable zone

Not enough heavy elements

Such regions favorable to life are called "habitable zones." In our own galaxy habitable zones are found in the calm "suburbia" that covers about 90% of the Galaxy, and our sun sits right in the middle of it. In the "inner city," the center of the Galaxy, the population density is too high, increasing the risk of destruction by radiation or collision. Out in "exurbia," the recycling rate for matter by supernovae is too low to create enough heavy elements.

Zooming in on individual stars, the habitable zone around them is the region where the surface temperature of a potential planet would allow liquid water to exist. The temperature at the surface of a planet depends on two factors: its distance from the star and the density and composition of its atmosphere. The distance from the star controls the total energy received by the planet (the more distant, the less energy), and the characteristics of the atmosphere result either in cooling or warming by the greenhouse effect (Q. 92). In general, the more luminous the star, hence the hotter, the larger the habitable zone.

For our Sun, the habitable zone extends from 0.85 to 2 times the radius of the Earth's orbit. Venus is too close to the Sun to be in the zone, but Mars is in it. The temperature at the surface of Mars is now too low for liquid water to exist because it has lost most of its atmosphere (Q. 51), but liquid water was present on Mars in the past and would still be there today if Mars's gravity had been stronger (i.e. if the diameter of the planet had been larger).

179 Can planets exist around binary stars?

Double sunset on Tatouine – artist's view.

In the *Star Wars* saga, the Skywalker clan has its base on Tatouine, a desert planet orbiting around two suns. This was an exotic and intriguing idea for science fiction writers, but experts in celestial mechanics thought such a situation was improbable because they expected that the inevitable gravitational perturbations would prevent planets from forming.

Surprisingly, this does not seem to be the case. A planet has been found orbiting the binary star Gamma Cephei, and computer simulations have now shown that planets can form as easily around two stars as around one. For their orbits to be stable, however, they must either orbit around only one of the two stars or be located far enough away from both of them.

This is good news for the apparition of life because more than half of all stars belong to multiple systems.

180 What are the odds that other intelligent life exists in our galaxy?

First of all, what kind of intelligence do we have in mind? Monkeys and dogs are certainly intelligent, octopuses and dolphins too. But what we would really be interested in is making contact with beings with whom we could communicate, share thoughts and information. This means that they would have to be self-conscious, have a language, be organized, and possibly have advanced technology; in other words, have developed a *civilization*.

Since we cannot simply go to meet those interesting beings face to face (assuming they have faces), as Christopher Columbus did when he sailed to America, we will have to restrict ourselves to advanced civilizations, i.e. those that have developed the means to communicate across galactic distances. Note that if we applied this same restriction to civilizations on Earth, we would be eliminating the civilizations of the Sumerians, of the Mayas and Incas, of Shakespeare and Beethoven, and certainly those of our prehistoric but fully evolved Homo sapiens ancestors. A pity, but on the scale of evolutionary time, millions of years, those 50 000 years between the appearance of modern man and the invention of radio communication less than 100 years ago is nothing, just the blink of an eye.

In 1961, when the American radioastronomer Frank Drake was preparing the first conference on radio detection of extraterrestrial intelligence, he decided to estimate the chances of finding other civilizations in the Galaxy as a way of convincing participants that those chances were rather better than zero. The result was the famous equation that now bears his name:

Frank Drake.

$$N = R_\star \times f_p \times n_e \times f_l \times f_i \times f_c \times L$$

in which,

N = number of civilizations in the Galaxy with which we could communicate,
$R_\star$ = rate of formation of stars similar to the Sun,
f_p = fraction of these stars having planets,
n_e = number of planets per star that can harbor life,
f_l = fraction of those planets where life does appear,
f_i = fraction of those where intelligent life develops,
f_c = fraction of those where intelligent life is advanced enough to be able to communicate across space (i.e. advanced civilization),
L = duration of these civilizations.

This equation has withstood the test of time. Unfortunately, however, only the first two terms can be estimated with relative certainty: $R_\star$ is of the order of 3 stars per year (Q. 30)[†] and, according to the latest exoplanet findings, f_p is of the order of 0.5. Beyond these two terms, uncertainty prevails. Let us go ahead all the same and set $n_e = 1$ and $f_l = 0.5$. As for f_i, the chance that intelligent life will evolve, biologists are divided. For some, evolution advances at random, without direction, and intelligence can only appear due to an extraordinary series of chance events. For others, intelligence is such a formidable survival trait, increasing the chance to find food and protection, that its appearance is inevitable. Let us opt for $f_i = 0.5$ and $f_c = 0.3$; we then have approximately $N = L/10$. The duration of the advanced civilizations, L, is doubtless the most difficult term to estimate. Applied to our own civilization, if we were to exterminate ourselves soon in a nuclear war, L would only be 60 years (from the time of the first radio telescopes in the 1950s to the present)! If we can avoid destroying ourselves and overexploiting our planet, our civilization and subsequent ones might see our species millions of years into the future, maybe even until the death of our Sun 4 billion years from now. If we are not too pessimistic about the capacity of civilizations to preserve themselves, adopting $L = 10\,000$ years, the number N of civilizations in the Galaxy with which we could communicate would be of the order of 1000. If advanced civilizations can be trusted to preserve themselves over, say, 10 million years, then N would be a million.

But maybe you see the probabilities differently. Work it through yourself, then, using your own estimates. Assign a value to each term. After that, just six little multiplications – no logarithms, no trig., no calculus. If you find a high number for N, you will doubtless become an enthusiastic supporter of SETI (Q. 186). Still, even if the number you get is small, wouldn't it still be tempting to listen to the cosmos . . . just in case?

181 Where else in the Solar System could life exist?

For a while, in the late nineteenth and early twentieth centuries, several astronomers were convinced that they were seeing canals on Mars. Some even went so far as to propose that they were irrigation canals built by an intelligent civilization. More careful observations did not confirm the discovery, however, and spectroscopic analysis also began to reveal the absence of water in the Martian atmosphere. Since then, all the planets and their satellites have been observed – even explored – thoroughly enough to conclude without a doubt that there is no intelligent life in the solar system outside of Earth. But what about primitive forms of life?

Liquid water seems to be essential for life (Q. 174). No hope then for Mercury or Venus, where surface temperatures are over 450 °C. No hope for the Moon, either, which has no liquid water, no atmosphere, and where the lack of life has been confirmed by numerous experiments in situ.

The giant planets – Jupiter, Saturn, Uranus, Neptune – are gaseous, with no solid external surface and without liquid water. Comets do contain water, and so probably

---

[†] Only F, G, K, and M type stars (Q. 13) are taken into account. O and B stars have too short a lifetime for life to develop.

Cliff formations on Mars which were probably carved out by floods of water.
Photo taken by the NASA Mars Express probe in 2008.

do some Kuiper Belt objects, but only in the form of ice, not as a liquid. That leaves
Mars and some satellites of the giant planets.

The average temperature of the Mars surface is −60 °C. Temperatures do exceed 0°C
at the equator during the Martian summer, but since the atmospheric pressure is only
1% of that of the Earth, water cannot stay liquid, it "sublimates" (i.e. goes directly
from ice to vapor). Several high-resolution photos of the Martian surface taken by
probes have revealed signs of water erosion, however. This could mean that liquid water
was present at earlier times, when the Martian atmosphere was denser and atmospheric
pressure therefore higher. Pockets of liquid water could still remain deep under the
surface where microorganisms might survive. Maybe one day we will find fossils of
organisms dating back to the era when the surface of Mars had liquid water.

Among the satellites of the giant planets, only Jupiter's Europa could harbor liquid
water. The others are either too cold or lack atmospheres. Europa, which is roughly the

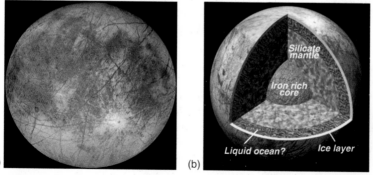

(a) Photo of Europa taken by NASA's Galileo probe in 1996. The bluish–white regions
represent ice, the brownish areas rocky material that emerged from the interior either
because of geologic activity or impact. (b) A schematic view of the interior of Europa.

size of our Moon, has a very smooth surface mostly made of ice. Its surface temperature is −160 °C at the equator, but intense tidal forces created by Jupiter keep its interior warm (Q. 52). The thick layer of ice at the surface likely covers a vast ocean which could be as deep as 50 km. Could living organisms exist there? Why not? Organic compounds may have been brought there by bombarding meteoroids. And although it would be far-fetched to expect life to have evolved there exactly as it did on Earth, producing fish with bony skeletons, scales, and gills, we might yet find some form of macroscopic life there, surviving without solar light, basking in the heat of deep thermal springs or currents.

182 How are exoplanets detected?

A planet orbiting a star other than the Sun is called an "extrasolar planet" or "exoplanet." The first exoplanets were discovered in the 1990s, and since then the harvest has been bountiful, with a new one being discovered almost every month. By late 2008, over 330 had been cataloged. Since planets are much smaller than stars and generally only shine by reflecting the light of their parent star, they shine very faintly. And besides being infinitely fainter than their parent star, they are angularly very close to them. A telescope is normally blinded by the light of the star so that, with a few exceptions (see figure), it is impossible to distinguish the planet. So how are they discovered?

   The most successful method is to observe the gravitational influence of the planet on its parent star. As in the case of the Moon orbiting around the Earth (Q. 107), or of stars in a binary system (Q. 8), a planet does not actually rotate around its parent star, but around their common center of gravity. And the star, of course, does likewise. If a star is seen to wobble over time in a periodic fashion, it is because one or several planets must be in orbit around it. By carefully analyzing the star's cyclic motion, then applying Kepler's laws (Q. 8), the mass of the planets and their orbital radii can be derived. In practice, the motion of a star in the sky is too small to be observed directly, but it can still be detected thanks to the Doppler effect (Q. 141): the light of the star becomes bluer when the star comes towards us, redder when it moves away. It is thanks to this method, called "radial velocity," that most of the exoplanets have been discovered so far. The disadvantage of this method is that it is only sensitive to heavy planets in close orbit around their parent stars. Heavy planets such as Jupiter or brown dwarfs (Q. 50)

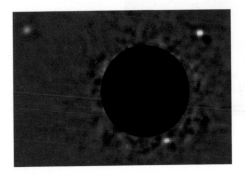

Near infrared image of three self-luminous giant Jovian planets orbiting the nearby star HR 8799 obtained with adaptive optics. The star has been masked for easier display and the three planets appear as the whitish dots. This planetary system, orbiting a larger and brighter star, seems to be a scaled up version of our solar system. Credit: C. Marois et al./Gemini Observatory.

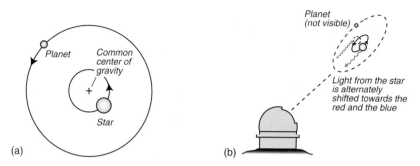

(a)　　　　　　　　　　　　　　　　　　(b)

(a) As a planet revolves around a star, the star rotates around their common center of gravity. (b) The motion of the star is detected using the Doppler effect.

can be detected, but not lighter planets such as the Earth because, there, the motion of the parent star is too small to detect.

For Earthlike planets, direct detection methods must be used. Capturing images of these faint planets is extremely difficult and would require observations from space using specialized instruments that are currently only on the drawing board (Q. 183). So, for the moment, we must fall back on quasi direct methods. One such method, called *transit*, is to measure the decrease in brightness of a star when a planet passes in front of it. This is currently being attempted from the ground on a relatively small number of stars. More promisingly, the space mission *Kepler*, which was launched in early 2009, will increase the chances of detection by simultaneously observing 100 000 stars.

Another quasi direct method consists of taking advantage of a gift of nature and of Albert Einstein: the *gravitational lens*. According to the theory of relativity, gravity deforms space. Consequently, when light rays pass close to a massive object such as a star or a planet, the rays are bent. And if a nearby star is almost aligned with a distant one, the light rays from the distant star will be bent and focused just as if the closer star were a giant lens. When this happens, the light of the distant star is greatly amplified. The amplification is transitory, just a blip in brightness, because stars seen from the Earth are always moving relative to each other due to the rotation of the Galaxy. The alignment typically lasts only a few hours. If the nearby star has a planet, its gravity

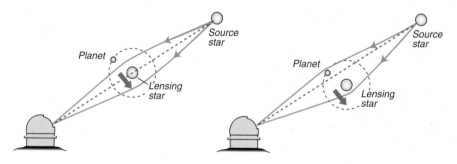

As a star passes between the observer and a more distant star, the phenomenon of the gravitational lens amplifies the brightness of the star in the background. A second brightness peak occurs if the closer star has a planet.

will create an additional image and a second blip in the brightness of the distant star. Such alignments are rare and a large number of stars must be maintained in continuous observation, but the method is sufficiently sensitive to detect planets of low mass such as the Earth.

183 How could we detect the presence of life outside the Solar System?

Imagine looking at Earth and the other planets from a position in space far outside the solar system. If we analyzed the light coming from each of the planets as a function of wavelength, Earth would stand out because its "spectrum" would have unique features.

The light coming from a planet is either reflected sunlight or is produced by its own heat (in the infrared). In either case, the light goes through the planet's atmosphere, and some of it is absorbed by the different gases it contains. Each type of gas absorbs light at characteristic wavelengths, giving it a distinctive "spectral signature." The presence of a given gas in a planetary atmosphere can thus be determined by analyzing the spectrum of the light received from that planet, then searching for its signature.

Observing in the infrared, we would detect water vapor and ozone in the atmosphere of the Earth, but not in that of Mars or Venus. Water and ozone are "biomarkers" indicating the potential presence of life: water, because it is eminently favorable to life; ozone, because it indicates the presence of oxygen, which can itself only be there if living organisms produce it. The reason for this is that oxygen is highly reactive, combining rapidly with many other chemical elements (in rocks and soil, for example), so that it can remain present as a gas only if constantly renewed. This renewal occurs

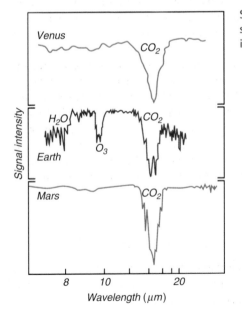

Spectra of Venus, Earth, and Mars. The signature of water (H_2O) and ozone (O_3) is only found in Earth's spectrum.

One of the proposed space missions for detecting life on exoplanets. Here, the European Space Agency's *Darwin*, composed of telescopes flying in formation. Credit: ESA.

through photosynthesis that is carried out by plants and marine phytoplankton (Q. 85). If life disappeared from Earth, the oxygen in the atmosphere would be eliminated after just a few thousand years.

Just as we could conclude from the spectra that there was life on Earth but very likely not on Mars or Venus, detecting the presence of water and ozone in an exoplanet's atmosphere would be a strong hint that it harbors life.

Methane, which on Earth is produced by bacteria and some animals (cattle, in particular) is also a good biomarker. Although methane can be non-biological in origin, it decomposes rapidly, particularly in the presence of oxygen. So, finding both methane and oxygen in a planet's atmosphere would indicate that the oxygen is being renewed biologically.

These are the principles but, in practice, analyzing atmospheres of exoplanets is enormously difficult: the light from the planet has to be separated from that of the nearby parent star which is between a million and a billion times more luminous. One approach is to observe in the infrared (where the star-to-planet ratio of luminosity is lower), from space (so that the telescope can be cooled, thus minimizing its own infrared radiation – see Q. 222), and with a long base interferometer to improve angular resolution and better separate the planet from the star (Q. 227). NASA and the European Space Agency (ESA) both have such projects on their drawing boards, with expected launches in the 2020s (Q. 223).

184 Could the human race ever colonize exoplanets?

The most interesting exoplanets for human colonization would naturally be those with a solid surface (i.e. a "rocky" rather than a gaseous planet), moderate temperatures, and capable of harboring human life – so probably containing liquid water.

The nearest habitable exoplanets that we can hope to discover are about 30 light-years away. With current technology, spaceships can reach a speed of 12 km/s, or 0.004% of the speed of light. At that rate, a round trip would take 750 000 years.

In order to make a round trip during a single human lifetime, we would need to travel at nearly the speed of light. Unfortunately, that is out of the question because of the enormous energy required. In particle accelerators, we have managed to accelerate protons to within 1% of the speed of light, but the mass of a proton is exceedingly small. Accelerating a 100 ton spaceship to within 99% of the speed of light would require 60 000 billion billion joules.[†] That turns out to be a fantastic amount of energy, equivalent to the worldwide consumption for 100 years (at the current rate, including energy from all sources).

So we are prisoners of Earth, or at least of the Solar System!

185 Could aliens have visited the Earth?

"You know — Once you add up the cost of fuel, food, and 110 years of babysitting, invading Earth hardly seems worth it ..."

Sensational UFO reports keep the tabloids happy, but none has ever withstood serious scientific investigation. There is no tangible evidence: no unquestionable photo, no sound recordings, no detectable strange residues, and not the tiniest fragment of an alien space ship. Surprising? Not in the least – because the laws of physics are the same everywhere in the Universe, and those laws that limit our own space travel also limit the space travel of alien civilizations.

Although extraterrestrial civilizations with technology far superior to ours may very well exist in the Universe, we Earthlings already know enough science to assert unequivocally that the basic laws of physics cannot be circumvented. No conceivable energy source could supply the gigantic amount of energy required for interstellar travel, for one thing. And for another, such journeys would take so long that the crew members would have to be near immortals to survive them.

Even then, one would need to know where to go; exploring the Galaxy at random would be an exercise in futility. If we make the optimistic assumption that one million advanced civilizations exist in our galaxy (Q. 180), and that, among those civilizations, one in a hundred decides to explore its stellar neighborhood, that would result in as many as 10 000 spaceships criss-crossing the Galaxy at any one time. Still, with 100 billion stars in the Galaxy, each spaceship would have 10 million stars to explore [22]. The chances that one of them would find a planet with intelligent life is ridiculously low.

[†] Calculated using relativistic kinetic energy $E = mc^2(1/\sqrt{1 - v^2/c^2} - 1)$, but we would also have to factor in the mass of the fuel required.

186 How could we communicate with other civilizations in the Galaxy?

If we cannot travel to even the nearest exo-planets (Q. 184) or count on a visit from little green men (Q. 185), could we at least hope to communicate somehow with other civilizations in the Universe?

Well, we could certainly let them know that we are here. And we could also listen to find out if somebody is out there. But communicate as if on a cell phone? Not quite. The nearest exoplanets are probably at least 30 light-years away, so the "dead time" between sending a question and receiving an answer would be at least 60 years ... Still, it would be a shame not to try.

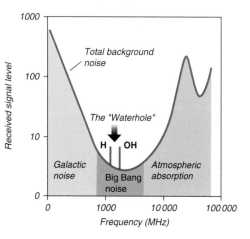

Any part of the electromagnetic spectrum could be used for communication – light for example – but radio waves are the best choice. Since stars do not radiate much at radio wavelength, we could detect a broadcast from a nearby planet (or have one of our own broadcasts detected) rather easily in that part of the spectrum. In addition, interstellar dust and gases absorb radio waves much less than they do light.

The first attempt to detect an alien broadcast was made in 1960, when the large radio telescope at Green Bank, Virginia was trained on two of the closest stars, Tau Ceti and Epsilon Eridani. No meaningful signal was detected. Since then, efforts to detect radio signals have been repeated and are still being performed under the auspices of the SETI program (Search for ExtraTerrestrial Intelligence). About 1000 stars similar to the Sun and located less than 200 LY away have been listened to. Still nothing.

Unfortunately, we do not know what frequency to tune into, so we scan the radio spectrum, usually concentrating our efforts on the band that would seem to be the most promising for interstellar radio communications, the band between 1000 and 10 000 MHz. That is where "static" due to cosmic emissions and atmospheric absorption is at a minimum. The hydrogen atom (H) produces strong radio emissions in that band which is prevalent in the Milky Way and in other spiral galaxies, but on either side of that emission line, good conditions prevail.

The hydroxyl molecule (OH) also emits in that band. Since H + OH = water, and water is essential to life, some people think that alien civilizations might attribute a special significance to the band between H and OH and choose the frequency band between these two emission lines to communicate across space. Dubbed the *waterhole*, a reference to the ponds in Africa where animals meet to drink in peace, the cosmic waterhole between H and OH might prove to be the peaceful meeting point of all civilizations in the Galaxy.

Listening for signals is all well and good, but sending is important, too. In 1974, the giant radiotelescope at Arecibo, Puerto Rico, transmitted our first message, the equivalent of a simple "here we are!" It was sent, once only, in the direction of the

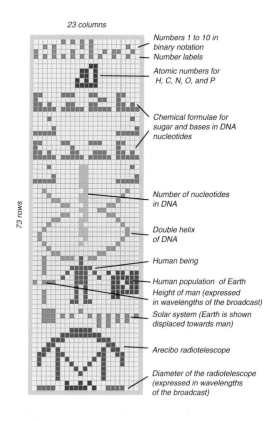

23 columns

73 rows

Numbers 1 to 10 in binary notation

Number labels

Atomic numbers for H, C, N, O, and P

Chemical formulae for sugar and bases in DNA nucleotides

Number of nucleotides in DNA

Double helix of DNA

Human being

Human population of Earth

Height of man (expressed in wavelengths of the broadcast)

Solar system (Earth is shown displaced towards man)

Arecibo radiotelescope

Diameter of the radiotelescope (expressed in wavelengths of the broadcast)

Message transmitted by the Arecibo radiotelescope in 1974 composed of a series of 1679 bits (0s and 1s expressed in two different wavelengths). No matter what type of counting system might be used by a civilization, a mathematician would recognize that 1679 is the product of two prime numbers, 23 and 73. When organized into 23 columns and 73 rows, the message reveals the image represented here. The white squares correspond to the 0s and the colored squares to the 1s. The color has been added here to help make the image stand out.

globular star cluster M13, which contains 190 000 stars. This multiplies our chance of being detected, but as M13 is 24 000 LY away, we cannot expect an answer for 48 000 years! A second message, repeated several times at each emission, was sent out in 1999, 2001, and 2003 by the 70 m radiotelescope at Eupatoria, Ukraine. This time it was aimed in the direction of several stars suspected of having Earthlike planets, and since these stars are only between 30 and 70 LY away, we could hope for an answer late this century or early in the next one.

Some people may deplore our failure to detect any messages so far and leap to the conclusion that we are probably alone in the Galaxy. On the other hand, what have we Earthlings really done to make our presence known since the invention of the radio beyond "leaking" radio waves into space? We have sent out only one purposeful *directed* message lasting just 1679 s, plus a few other signals totaling but a few hours ... Is that enough?

History of astronomy

187 Why did ancient astronomers study the sky so intently?

The Sun dictates our daily activities, and the change of seasons governs our agriculture and livestock management. From earliest times, our ancestors have watched the sky, derived the time of the day by the position of the Sun, and tracked it and other celestial bodies at night to predict the change of seasons and orient themselves on land and sea.

The ceremonial megalithic monument of Stonehenge (3000 BC).

It was evident to all that, without the Sun, life would wither and die. And soon, by extension, similar powers were attributed to the other "moving" bodies in the sky, the moons, planets, and comets. Our ancestors examined them all anxiously, trying to foretell events that could affect their destinies [7]. No wonder then that, between the need for calendars, navigational aids, and heavenly portents, the celestial vault was studied so early and so carefully.

Very early on, burgeoning astronomers in the Middle East, Egypt, and China measured the positions on the horizon where the Sun, Moon, and a few bright stars (Sirius and the Pleiades, in particular) rose and set. The wood or stone instruments at their disposal were rudimentary but precise enough to measure the positions to within an angle of half a degree (limited mainly by atmospheric refraction near the horizon – Q. 210). Mathematical tools (basic trigonometry and interpolation) were eventually developed to improve their calculations.

188 How did the cult of the Sun originate?

Our relationship with the Sun is a profound one, influencing not only our physical but also our emotional lives. A day without Sun makes us gloomy. Sunny beaches attract us like magnets. Sunsets and sunrises have universal appeal. It is easy to see how such a dominant celestial object could come to be deeply revered, especially when its nature and movements were so mysterious.

The Sun was honored as a god in most ancient civilizations. In Mesopotamia, the Sun-god was Shamash and for the Egyptians, Aton/Ra. The Greeks had Helios, the

Pharaoh Akhenaton (circa 1370 BC) making an offering to the Sun.

Romans, Apollo, and the Hindus, Surya. The Aztecs, who called themselves "the people of the Sun," performed cruel, sophisticated rituals of human sacrifice to feed their solar god Huizilopochitli and keep him – and the Universe – alive.

189 Why were the Greek and Roman gods associated with the different planets?

The first archaeological evidence for an association between the planets and the gods of antiquity is found in Sumer circa 2600 BC. The Sumerians were the first to record the irregular movements of the five planets visible to the naked eye: Mercury, Venus, Mars, Jupiter, and Saturn. Having concluded that all "wandering" celestial objects – the Sun, the Moon, and the five planets – were supernatural beings (Q. 187), they revered seven cosmic divinities: Utu, the Sun; Nanna (later Sin), the Moon; Enki, Mercury; Inanna,

A stela with the symbols of Mesopotamian deities: at top, from left to right, as if hanging from the celestial vault, are Sin, the crescent Moon, the goddess Ishtar, Venus, and the radiating shape of Shamash, the Sun god. Kuduru of Meli-shipak II at the Louvre Museum.

Venus; Nergal, Mars; Enlil, Jupiter; and Ki, Saturn. At the time of the Babylonians, the names of the gods were changed, the Sun god became Shamash, the Moon, Sin, and Venus, Ishtar. These planet-god associations were eventually retransmitted to the Egyptians, Greeks, and Romans (Q. 53).

190 Can we learn anything from the astronomical phenomena reported in the Bible?

The Bible cannot be considered a historical document any more than most ancient legends, and it is even less a treatise on science or astronomy. The few descriptions of astronomical phenomena in it are rudimentary and so vague that they contain little real information, unlike the documents of Mesopotamia and ancient China, which can be used to reconstruct the history of eclipses, for example.

The cosmogony of the book of *Genesis* is directly inherited from more ancient texts. The two biblical accounts of the birth of the Universe, the Creation Week and the Eden Narrative, are similar to and most likely based on Sumerian texts written around 2100 BC, more than a thousand years before their biblical rerun.

On the other hand, the Bible does contain some poetic texts that were clearly inspired by celestial objects or phenomena. God's response to Job's speeches is a fine example, in this description of the constellations (Job 38:31):

Can you fasten the harness of the Pleiades,
　or untie Orion's bands?
Can you guide the morning star season by season
　and show the Bear and its cubs which way to go?
Have you grasped the celestial laws?
　Could you make their writ run on the Earth?

191 How could the ancient astronomers predict eclipses?

Ancient astronomers could predict lunar eclipses well, but not solar eclipses. Lunar eclipses, the passing of Earth's shadow over the Moon, are rather frequent and can hardly be missed because they occur at night during a full Moon, and are visible from every point on Earth. Solar eclipses, on the other hand, are rare at any given location and are difficult to detect because the Sun is so bright. They go unnoticed unless they are nearly total or the Sun is veiled by clouds.

Babylonian astronomers began measuring the positions of the Moon and planets, night after night, in the second millennium BC. By around the fourth century BC they had compiled enough data to discern regular patterns and were able to begin predicting eclipses. Eventually they discovered the 18-year Saros cycle of lunar eclipses, when the Sun, the Earth, and the Moon return to approximately the same geometry. However, since they could not establish solar eclipse periodicities, they were limited to predicting their "chances" of occurring.

The early Chinese astronomers had a similar tradition of observations and calculations and hold the record for the oldest written reference to a solar eclipse (around 2000 BC).

In his *Almagest*, the Greek astronomer Ptolemy provides a list of lunar eclipses starting at 721 BC that agrees remarkably well with the Chinese list.

192 Who were the most important astronomers of antiquity?

Many of the ancient scholars who contributed to the progress of astronomy are unknown to us. The great majority of them, in particular the Babylonian and Egyptian pioneers, will forever remain in the shadows. The few classical Greek astronomers listed below would never have achieved what they did without the foundations laid by their many anonymous predecessors and contemporaries.

Aristarchus of Samos (circa 320–250 BC), known as the Copernicus of Antiquity, proposed the revolutionary and heretical concept for his time that the Earth and planets revolve around the Sun. He maintained that the stars were other suns, immensely far away, accounting for their faintness and absence of parallax (Q. 6). Using rigorous trigonometric methods, he was the first to calculate the distance to the Moon and the Sun. But measuring the enormous distance to the Sun required better instruments than he had available, and he underestimated it by a large factor. Still, his method was fundamentally correct.

Eratosthenes of Alexandria (circa 276–194 BC), Greek astronomer, mathematician, and geographer, is one of the most famous scientists of all times. He was the first to measure the size of the Earth (Q. 79).

Hipparchus of Nicea (circa 190–120 BC) is another shining light among the great figures of ancient astronomy. He compiled an extensive star catalog (Q. 5) that was used for almost 2000 years to identify novae and supernovae. A serendipitous result of his careful measurements was his discovery of the precession of the equinoxes (Q. 87), which he estimated at 1°/century (the modern value is 1.38°). Hipparchus also proposed using lunar eclipses to determine geographic longitude.

Claudius Ptolemy of Alexandria (100–165 AD) is the author of a comprehensive work summarizing classical Greek astronomy, the *Greatest Compilation*, better known by its Arabic title, the *Almagest*. This monumental work in 13 volumes became the basic reference work for Islamic and medieval scholars. Ptolemy also made original contributions to the science with his precise determination of the distance to the Moon, using Aristarchus's parallax method, which he significantly improved. However, his reputation and influence are mainly due to his famous geocentric model of the Sun and planets (Q. 194). This was a complex model that eventually proved to be wrong, but it did predict the positions and movements of celestial bodies with much greater accuracy than had been achieved previously.

The Greek civilization disappeared, and the western world was plunged into an intellectual night that lasted 1000 years. Only the strong Arabic and Byzantine respect for ancient science kept the astronomical and geographical knowledge of antiquity alive.

193 What were the contributions of the Chinese, Indian, and Islamic civilizations to astronomy?

In view of the fundamental contributions made by Mesopotamia, Egypt, Greece and, beginning in the seventeenth century, by Europe and America, one might be tempted to conclude that astronomy is overwhelmingly a western science. Were the two great civilizations of the East, India and China, left behind? If not, what contributions did they make?

Indian astronomy has ancient roots, going back as early as 2000 BC in the Indus Valley. It was then probably centered on establishing calendars. Scientific astronomy only began to develop in India around the sixth century AD after coming into contact with the Greek and Byzantine traditions. At that time, Indian mathematics was already well advanced, probably more so than anywhere else in the world, and astronomers used that to advantage. Aryabhata (476–550 AD)[†] and Brahmagupta (598–668 AD), the two dominant mathematician-astronomers of this epoch, used new mathematical techniques to establish tables showing the positions and movements of planets, the phases of the Moon, and solar eclipses.

In the eighteenth century, Maharajah Sawau Jai Singh, an astronomy enthusiast, had several observatories built, including the Jaipur Observatory which is now a tourist attraction. These observatories were equipped with large instruments used for measuring the positions of stars and planets with great accuracy. It was a medieval kind of astronomy, however, contributing no new knowledge, whereas a century earlier Galileo, Copernicus, and Newton had already transformed the science.

Chinese astronomy, also very ancient, had its beginnings as early as 2000 BC and developed quite independently. It was at least as advanced as western astronomy when it came into contact with European civilization in the sixteenth century.

Chinese science generally differs from its western counterpart by its very practical approach. While ancient Greek scholars were fascinated by theoretical questions and felt that practical applications were

Jaipur Observatory (1734).

[†] Aryabhata taught that the Earth rotated while the stars were fixed, and it has been alleged that he actually proposed a heliocentric model, but there is no firm evidence for this.

beneath them, in China it was the reverse. The main purpose of Chinese astronomy was not intellectual investigation, but calendars and divination. The Emperor embodied the connection between human beings and nature, and the calendar was a manifestation of this relationship. Astronomers had to create new calendars based on new observations every time a change of dynasty took place.

For the Chinese the sky was also an open book that held the secrets of the future. It was important to learn how to read and interpret it. Whether for the Emperor's needs or common people's daily lives – a businessman's prospects, choice of a spouse, destiny of a newborn – it was important to scrutinize the sky, which could provide useful clues.

These sociopolitical connections may explain why Chinese astronomers kept detailed records of astronomical events from very early on, leaving us lists of a thousand eclipses, 360 comet passages (31 of which refer to Halley's comet), 700 meteoric falls, many novae and supernovae, including the famous supernova of 1054 in the Crab Nebula, and lists of sunspots from as early as 28 BC.

Ferdinand Verbiest (1623–88), a Jesuit priest, dressed as a Chinese astronomer.

These catalogs, covering almost 3000 years, have been a gold mine for modern astronomers. Yet, although Chinese astronomers could predict eclipses and realized that planets were different from stars, they did not attempt to *understand* what they were observing. For this reason, in spite of its relative advance, Chinese astronomy did not bring much to western astronomy once it had renewed contact with Europeans in the sixteenth century. An interesting anecdote illustrates the superiority of western astronomy by then. In 1668, Rome sent a small group of Jesuits, including the Flemish astronomer, Ferdinand Verbiest, on a mission to China. Wishing to compare the relative strengths of Chinese and European astronomy, Emperor K'ang Ksi organized a contest involving three predictions where "the heavens would be the judge." The two contestants had to predict: the length of the shadow of a sundial's gnomon at noon, the positions of the Sun and planets on a given date, and the exact time of an upcoming lunar eclipse. The Chinese astronomer was wrong on most counts, while Verbiest's answers were all perfect. As a result, the Emperor appointed the Jesuit father head of the Chinese Imperial Bureau of Mathematics and director of his observatory.

On the other hand, Islamic scholarship and civilization played a crucial role in the preservation of ancient knowledge and transmission to the western world of Greek, Babylonian, and Indian science, and to a lesser degree, of Chinese achievements. The ancient texts were translated into Arabic in the wake of the lightning conquests of Islam during the seventh and eighth centuries, and passed on to the Christian world of Europe via northern Spain starting in the eleventh century.

Three great figures of Islamic astronomy: from left to right, the Syrian Al-Battani (868–928), a meticulous observer who accurately determined the duration of the year and discovered what is now called the ellipticity of the Earth's orbit; Averroes, or Ibn-Rushid (1126–98), from the famous school of Andalusia, who developed a reasoned critique of Ptolemy's geocentric model; and the Persian al-Tusi (1201–74), who refined Ptolemy's geocentric model based on more accurate observations.

Islamic astronomers and mathematicians (many of them Persian) also made contributions of their own. They built the first true observatories (Q. 196), refined the astronomical measurements of antiquity (length of the year, value of the precession of the equinoxes, tables of positions for the Sun and the Moon), and developed new methods of calculation based on trigonometry and spherical geometry.

Islamic astronomy provided a solid base for the development of astronomy in medieval Europe, for the birth of Copernicus's heliocentric theory, and for the development of the new instruments and techniques that were to be so successfully used in Europe in the sixteenth and seventeenth centuries.

194 Who was responsible for overturning the geocentric system?

The Earth feels steady under our feet, appears to be flat, and the sky seems to revolve around us. Perhaps it is not surprising, then, if that evidence plus a generous serving of vanity pushed human beings into imagining themselves at the center of the Universe. That we eventually managed to prove that our planet is a rotating sphere floating in space and in orbit around the Sun is one of the most important victories of rational thinking over "common sense" ever achieved by the human race. How did this happen?

The ancient Babylonian and Egyptian astronomers may have been keen observers of the heavens, but they worked empirically, measuring the repetitive movements of the Sun and the planets in order to predict their future positions. Happily, first in Babylon, then in Greece in the sixth century BC, there emerged a new way of looking at the world that was to change the course of human thought: rational thinking replaced myth and superstition as a way of comprehending the world.

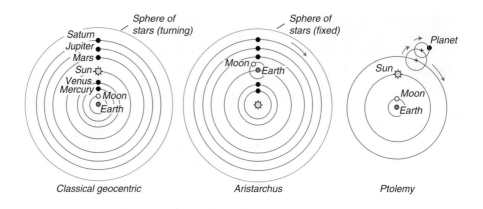

Classical geocentric   Aristarchus   Ptolemy

As Greek philosophers and astronomers struggled to comprehend the reasons behind the movements of the errant celestial bodies, they were obliged to develop a mechanistic "system" – what today we would call a "theoretical model" – and a way to confront this model with their observations. Bit by bit they convinced themselves that the Earth was a sphere and suspended in space, with the Sun, planets, and fixed stars revolving around it in a merry-go-round configuration. To explain the motions of the Sun and planets against the background of fixed stars, they positioned them on orbits of different radius. But there was a problem: observations clearly showed that while the Sun and the Moon moved in a regular and continuous manner, the five planets showed erratic and discontinuous behavior. Around 280 BC, Aristarchus of Samos proposed putting the Sun at the center of the Universe and having the planets, including Earth, orbit around it (Q. 192), which would have solved most of the observed "irregularities." Unfortunately, Aristarchus left no disciples, and his heliocentric model was abandoned because it contradicted the ideas of the great philosopher Plato.

Plato was not a scientist: science, particularly astronomy, bored him. Based on his own metaphysical reasoning, he concluded that the Universe had to be a perfect sphere and that the motions of the seven "errant" celestial bodies were on circular orbits moving at uniform speeds [27]. His disciple Aristotle adopted this view and subsequent generations of astronomers wrestled with the need to fulfill these preconceived conditions. This led to the monster of complexity that is the geocentric system of Ptolemy, where the erratic motions of the planets were explained by having them move on nested sets of circles ("epicycles"). When more precise observations did not match the model, additional epicycles were introduced or the positions of the centers of the orbits were fiddled with.

This outlandish situation lasted for 1200 years. A number of fifteenth-century scholars and astronomers showed the way, but it was a quintet of astronomers and physicists of the sixteenth and seventeenth centuries that demolished the Aristotle–Ptolemy system once and for all.

The first actor to come on stage was the Polish canon **Nicholas Copernicus** (1473–1543). After analyzing a massive number of measurements, he demonstrated that one could dispense with many epicycles simply by having the planets, including the Earth, go around the Sun. He explained the diurnal motion of celestial objects by the rotation of the Earth on its axis, and the apparent fixity of the stars by their great distances. Nevertheless, he retained the idea of the planets moving in circular orbits at uniform speed, which forced him to preserve a few epicycles. His treatise, *De Revolutionibus Orbium Coelestium* (*On the Revolution of the Celestial Spheres*), was completed around 1530, but he delayed publication, fearing the wrath of the Roman Catholic Church. Published in Nuremberg in the year of his death, the work had a huge impact and triggered what is now known as the scientific revolution.

Danish astronomer **Tycho Brahe** (1546–1601) refused to accept Copernicus's heliocentric model. He preferred a hybrid model with the Sun and Moon in orbit around the Earth, and the planets revolving around the Sun. He was a remarkable maker of instruments (these were traditional sighting instruments – the telescope had not yet been invented), which he installed in his observatory in Uraniborg (Q. 196). With them he patiently measured the positions of the planets over several years with unsurpassed precision (on the order of one minute of arc), and these sets of data, in particular for Mars, became the basis for the revolutionary model of the solar system that would soon make its appearance.

It was the German mathematician astronomer (and astrologer at times) **Johannes Kepler** (1571–1630) who finally discovered the laws of planetary motion. Copernicus's model was able to explain the movements of Venus, Jupiter, and Saturn relatively well, but not that of Mars. Kepler tried all kinds of combinations of epicycles, off-set centers, and circles, but the errors remained as large as a degree, far too large compared to the exquisitely precise measurements of Tycho Brahe. He then decided to abandon the circle for an oval and, after searching for 7 years, finally found the solution: the orbits were *ellipses*, not circles, as astronomers had been assuming for the previous 2000 years. He published his results (now called the first two laws of Kepler, the law of ellipses and the law of equal area in equal time) in his treatise *Astronomia Nova* in 1609. A few more years of work and he had discovered the law governing orbital periods, now called Kepler's third law.

Kepler's discoveries supporting the Copernican model remained the province of scholars. The general public was unaware of them and the Catholic Church, which adhered to the Aristotelian doctrine wherein the Earth was at the center of a perfect Universe, was determined to keep things that way. The need to abandon Aristotle's model soon became evident, however. In 1609, an ebullient Italian physicist-astronomer, **Galileo Galilei** (1564–1642), directed his newly fabricated telescope at the sky. The Sun showed spots and the Moon, mountains! – that was the end of Aristotle's "perfect" Universe with unblemished celestial bodies. The sunspots moved along tracks that could best be explained by a rotation of the Sun on its axis combined with a revolution of the Earth around it – a major blow to the geocentric system. And Jupiter had moons – demolishing the belief that the Earth was the unique center of rotation in the Universe. Finally, Venus had phases which could only be explained by its orbiting around the Sun, not around the Earth ... Galileo's discoveries shook the educated world and led to his disastrous confrontation with the Catholic Church[†] which ended in prosecution by the Inquisition in 1633. Unfortunately, at the time of the trial, his strongest argument in favor of heliocentrism, the inclination and shape of the sunspot tracks, could not be verified because sunspots were entering the deep Maunder Minimum (Q. 42). In 1633 he was judged guilty of heresy and spent the rest of his life under house arrest. His condemnation was officially revoked by Rome only in 1992.

Although Kepler's laws accurately described the heliocentric trajectories of planets, the deep reason behind their movements remained unexplained. This was to be the major contribution of the English physicist **Isaac Newton** (1642–1727). In his brilliant work published in 1687, *Philosophiae Naturalis Principia Mathematica*, Newton showed that the force that keeps the planets in their orbits is the same force that maintains the Moon in orbit around the Earth, and the same that makes an apple fall: gravity. And he used the three laws of Kepler to prove that the force of mutual attraction between two bodies is inversely proportional to the square of the distance between them (Q. 80).

195 Who was the first astronomer to use a telescope?

Galileo is traditionally credited with being the first to look at the sky with a telescope – this, in early December 1609, soon after he had heard about the new Dutch invention and built his own instrument (Q. 214). But recent evidence has shown that an Englishman, Thomas Harriot, actually preceded him by a few months. Using an imported Dutch instrument, he made a map of the Moon on July 26 of that same year [12].

Harriot left no mark on astronomy, however, because he made few observations and did not publish any of his findings. Galileo, on the other hand, enthusiastically threw himself into an extensive exploration of the sky, told the whole world about it, and triggered a revolution in science.

[†] The newly formed Protestant churches and their leaders, Calvin and Luther, were no less against the Copernicus system than the Mother Church, in as much as they insisted on a literal interpretation of the Bible.

His telescopes were rudimentary, but the discoveries he made were spectacular: Jupiter had satellites; Venus had phases; the Moon was cratered; the Sun had spots; the Milky Way contained countless stars invisible to the naked eye.

Galileo quickly published his observations in a short treatise, *Sidereus Nuncius* (Starry Messenger), written in Latin and printed in March 1610, a work that spread like wildfire through the scholarly community of Europe [24].

Galileo's discoveries immediately stimulated a myriad of amateurs to equip themselves with telescopes to observe the sky. By around 1650, the Moon had already been mapped in great detail.

The Moon viewed by Galileo.

196 Where were the earliest observatories?

The astronomical observatory, in the sense of a scientific institution with a director, a team of observers, a library, quality instruments, and a research program, is an invention of the medieval Arab world.

The first Islamic observatory was probably built in the region of Ispahan, in what is now Iran, around 1074. Observations were conducted there over a period of 30 years in order to cover one complete orbit of Saturn [31]. Next came the great observatory of Maragha, near Tabriz, also in present-day Iran. Built around 1260 for the famous astronomer al-Tusi (Q. 193) by Houlagou Khan, the grandson of Gengis Khan, this observatory was remarkable for the quality of its instruments, its huge library (400 000 manuscripts) and the large number of astronomers who worked there. The observatory at Maragha served as a model for the great observatories to come, notably one in Samarkand (in present-day Uzbekistan), founded in 1420, and another in Istanbul, founded in 1575. These were great centers of learning that played a large part in the flowering of Islamic science.

The first European observatory was founded by Tycho Brahe on the small island of Hven, near Copenhagen. Built in 1576 with royal funds, Uraniborg, "the castle of Urania," (the muse of astronomy) housed a library, a workshop, a printing press, and several instruments including a large mural quadrant (left). It became a magnet in its day for the best minds of Europe. But, at the death of his royal benefactor, Tycho Brahe fell out of favor and had to flee to Prague (where he was to become the mentor of Johannes Kepler), and that was the end of the observatory. Although shortlived, Uraniborg served as an institutional model for the great royal observatories of Paris and Greenwich. It has been partially restored and can be visited today.

197 How did the modern observatory evolve?

Until the invention of the telescope, astronomical observatories were dedicated to the precise measurement of the position of stars and planets using sighting instruments (Q. 196). And the first observatories in western Europe built to house telescopes were not even founded to do astronomy per se, but principally to aid navigation. Commerce, exploration, and war created a desperate need for precise ephemerides and ways of finding longitude at sea. This led to the creation of the Paris Observatory in 1671 and the Royal Greenwich Observatory near London in 1675.

Observatories for astronomical research appeared only in the eighteenth century. Usually associated with universities, they could be found in a number of European cities: Berlin, Leiden, Uppsala, Stockholm, Lund, Vienna, Turin, Milan. The nineteenth century saw a renewed flurry of observatory construction, both associated with universities and privately funded. In the USA, the first serious observatory, the US Naval Observatory in Washington, DC, was built in 1842. They became a common addition to university campuses after the Civil War.

All of these observatories were typically built near cities and suffered from poor image quality due to light pollution and atmospheric turbulence (Qs. 213 and 218). It was the American astronomers at the dawn of the twentieth century, particularly George Hale and Edwin Hubble, who first chose sites for their large new telescopes on the basis of image quality rather than proximity to their offices. Thanks to their giant instruments and good sites in California, American astronomers quickly made huge strides. Astronomy, which until then had been essentially a European science, became an American one.[†] Half a century would pass before continental Europe recovered, with the construction of large telescopes in Chile and Hawaii (Q. 221).

198 What have the major milestones been in our quest to understand the Universe?

The main steps that have led to our present understanding of the Universe are summarized in the table below. It is important to remember, however, that, although the astronomers and physicists mentioned there pioneered important new ways of thinking, others before them had prepared the way, and still others followed, refining their findings. In the words of Isaac Newton: "If I have seen further it is only by standing on the shoulders of Giants."

[†] With the exception of Great Britain which was well served by its theoreticians and by its expertise in radioastronomy, which is unaffected by atmospheric turbulence.

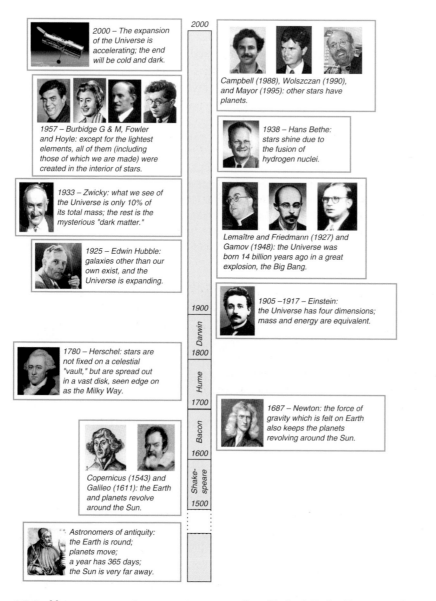

2000 – The expansion of the Universe is accelerating; the end will be cold and dark.

Campbell (1988), Wolszczan (1990), and Mayor (1995): other stars have planets.

1957 – Burbidge G & M, Fowler and Hoyle: except for the lightest elements, all of them (including those of which we are made) were created in the interior of stars.

1938 – Hans Bethe: stars shine due to the fusion of hydrogen nuclei.

1933 – Zwicky: what we see of the Universe is only 10% of its total mass; the rest is the mysterious "dark matter."

Lemaître and Friedmann (1927) and Gamov (1948): the Universe was born 14 billion years ago in a great explosion, the Big Bang.

1925 – Edwin Hubble: galaxies other than our own exist, and the Universe is expanding.

1905 –1917 – Einstein: the Universe has four dimensions; mass and energy are equivalent.

1780 – Herschel: stars are not fixed on a celestial "vault," but are spread out in a vast disk, seen edge on as the Milky Way.

1687 – Newton: the force of gravity which is felt on Earth also keeps the planets revolving around the Sun.

Copernicus (1543) and Galileo (1611): the Earth and planets revolve around the Sun.

Astronomers of antiquity: the Earth is round; planets move; a year has 365 days; the Sun is very far away.

199 Have any astronomers won the Nobel Prize?

Astronomy is not one of the five disciplines listed in Alfred Nobel's will: physics, chemistry, physiology and medicine, and literature (a "Nobel for economy" was added in 1969). As a result Edwin Hubble, who might well have deserved the recognition, never received it.

But astronomy has evolved into astrophysics, blurring the distinction between the two sciences, and several Nobel Prizes in physics have been awarded for astronomical discoveries or research in the last few decades. The first time was in 1974. That year, the physics prize was shared by two British radio astronomers: Sir Martin Ryle for his invention of the technique of aperture synthesis (Q. 227), and Anthony Hewish for the discovery of pulsars.

Since then, the Physics Nobel Prize has been awarded to several other astronomers:

- Arno Penzias and Robert Woodrow Wilson (USA) in 1978 for their discovery of the cosmic microwave background (Q. 133);
- Subrahmanyan Chandrasekhar (USA/India) and William Alfred Fowler (USA) in 1983 for their work on nuclear reactions in stars;
- Russell Hulse and Joseph Taylor (USA) in 1993 for their discovery of a binary pulsar and the demonstration of the effect of gravitational waves;
- Raymond Davis (USA) and Masatoshi Koshiba (Japan) in 2002 for the detection of cosmic neutrinos, and Riccardo Giacconi (USA), also in 2002, for his role in the development of x-ray astronomy;
- John Mather and George Smoot (USA) in 2006 for their measurement of the cosmic microwave background.

Without having the prestige of the Nobel Prize, the Bruce Medal, awarded yearly since 1898 by the Astronomical Society of the Pacific, recognizes the careers of researchers who have made fundamental contributions to astronomy. All the great names of twentieth-century astronomy have been among the recipients, including Henri Poincaré (France) in 1911, George Ellery Hale (USA) in 1916, Arthur Stanley Eddington (UK) in 1924, William de Sitter (The Netherlands) in 1931, Yakov B. Zel'dovich (Russia) in 1983, Vera C. Rubin (USA) in 2003, and Sidney van den Bergh (Canada) in 2008.

200 Astrology, astronomy, astrophysics ... what are the differences?

Astronomy is the science that deals with celestial objects. The term comes from the Greek *astron*, star, and *nomos*, laws.

Astrophysics is a branch of astronomy that arose at the end of the nineteenth century when astronomers became interested in the physical properties of celestial bodies (temperature, radiation characteristics, chemical composition, etc.), rather than just their positions and motions, and adopted some of the methods of physics, notably spectroscopy. Physicists, for their part, began using the sky as a laboratory to test their theories. An example of this synergy was a discovery made by two astronomers, the Frenchman Pierre Janssen and the Englishman Joseph Lockyer. Both were independently making spectroscopic observations of the 1868 total solar eclipse when they found a mysterious set of lines corresponding to no known element. The new element was given the name helium, from the Greek *helios*, the Sun. Another such example is the emergence of the theory of relativity, whose inspiration and applications are drawn as much from physics as from astronomy (Q. 159).

Modern astronomy has essentially become astrophysics, and the two terms are used almost indiscriminately at present. However, the term astronomy, although it actually encompasses astrophysics, can be used in the restricted sense to mean the study of the positions and movements of celestial objects.

An important branch of astronomy is **cosmology**, which is the study of the origin and structure of the Universe on the largest scale.

Astrology, despite having an etymology similar to astronomy's (*logos* means "the study of"), is not a science, but a practice based on the fanciful belief that events and human affairs can be influenced by the relative positions of celestial bodies. Astrology has long had an association with astronomy. In China, Babylonia, Rome, and in medieval Europe, many astronomers also held positions as astrologers.[†] Kepler and even Galileo practiced it. The final split took place in the seventeenth century, when the scientific revolution swept all irrational thinking out of science.

201 Is astronomy a "useful" science?

What causes the seasons? Where did the Earth and the Universe come from, and what is their destiny? These are some of the questions that astronomy addresses to satisfy our curiosity. But beyond providing intellectual satisfaction, is astronomy really useful? Useful like physics, that feeds all of modern technology, or like biology, that brings progress in medicine? The answer is a resounding yes.

First, it is good to remember that we do not live solely to *survive*, but also to take pleasure, and that pleasure is itself a survival factor, a source of energy that helps us to keep our mental balance. From this perspective, by answering many of the "Big Questions" about the cosmos, astronomy is as useful as music, theater, cinema, and all forms of art that profoundly satisfy us and enrich our lives.

In terms of pure practicality, astronomy actually started out as an applied science, not as an intellectual pursuit: its main purpose was to establish calendars, determine the time of day, and aid navigation. As it came into its own, it engendered the scientific revolution: it was thanks to the works of Kepler, Galileo, and Newton that physics took its first big steps and the power of scientific reasoning became overwhelmingly clear.

Astronomy is at the origin of a number of important mathematical tools (trigonometry, differential calculus, logarithms) and of developments in many branches of technology (optics, spectroscopy, photon detection, cryogenics). The symbiosis between astronomy and other scientific disciplines and technologies is now so intimate that many would certainly suffer if astronomical research were suddenly to stop. And the number of recent practical spin-offs from the science are countless, among them the detection of solar eruptions that affect our communications, the identification and tracking of asteroids and comets that may threaten Earth, and the development of certain imaging techniques used in medicine and manufacturing.

More importantly, perhaps, the study of the atmospheres of other planets has aided our understanding of our own atmosphere, and the discovery of the runaway conditions that devastated our sister planet, Venus, now serves to remind us of the fragility of our planet. This has helped to increase public awareness and, hopefully, may influence public policy as we face our current grave environmental challenges.

[†] This was not true of classical Greek astronomers, who viewed science as the reserve of pure mental activity, and felt that putting it to the service of ordinary human concerns would have been unworthy.

Astronomy looks outward, at the Universe, but it also helps us to look inward, putting our human wars, fears, aspirations, collective noble gestures, and base cruelties into perspective. It helps us to see ourselves as we are in that famous photograph taken by Voyager 1 in 1990, showing Earth from a distance of approximately 3 billion km, "a pale blue dot," in the words of Carl Sagan, "a mote of dust suspended in a sunbeam." If astronomy did no more than occasionally remind us in that way of how immensely privileged and immensely fragile is the small planet that we live on, it would more than have earned its right to be called "useful."

Telescopes

Astronomy is essentially a *passive* science. Aside from exploring the Moon and our nearest planetary neighbors, we cannot make experiments directly, we can only observe and try to understand what we see. And the master tool for making observations is the telescope.

202 How do refracting and reflecting telescopes differ?

Generally speaking, a telescope[†] is an instrument that enhances the observation of celestial objects by increasing their apparent size and luminosity. This applies to the entire electromagnetic spectrum, from radio waves to gamma rays, including of course the "optical domain," which covers visible light and radiation in the infrared and ultraviolet. Optical telescopes can be very diverse, but basically they work just like a photographic camera: they focus the light of a celestial object to form a *real image*[‡] on film or on an electronic detector. In the visual version, an "eyepiece," which is basically a magnifying glass, is used to observe the image directly.

The terms "refracting" and "reflecting" simply refer to the composition of a telescope's optics. If lenses are used, the instrument is a *refracting* telescope; if mirrors are used, it is a *reflecting* telescope.

Which system is better? Well, in general, the larger the telescope – that is, the larger its main mirror or lens – the more light it can collect and the fainter the object it can observe (Q. 205). Reflecting telescopes can be built very large indeed, whereas refracting telescopes have serious size limitations. This is because in a refractor the light must pass unobstructed through lenses, and hence they can be supported only at their rims. The problem is that a lens with a diameter of much more than 1 m sags under its own weight and lens deformation quickly becomes optically unacceptable. Reflecting

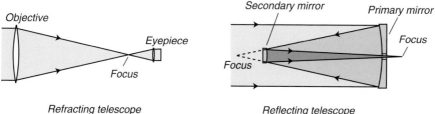

Refracting telescope   *Reflecting telescope*

[†] The term telescope comes from the Greek *tele*, "far" and *skopein*, "look."

[‡] That is, an image that can be projected onto a screen, as opposed to a *virtual image* such as the image of yourself in a mirror.

telescopes, on the other hand, have no such problems because their mirrors reflect light from their front surfaces. The mirrors can thus be supported over the entire back surface, preventing deformation of the optical surface, and can be built to very large diameters.

A second advantage of reflecting telescopes is their compactness. A second mirror can be used to fold the light beam back over itself, cutting the overall length of the telescope by approximately half. The largest refracting telescopes ever built are immensely long. The Yerkes Observatory refractor, which dates back to 1897, is 20 m long for a diameter of only 1 m.

A final advantage of reflecting telescopes is that they are by nature "achromatic": their images do not exhibit color effects. Unlike ordinary mirrors which reflect light from their backs, astronomical mirrors are coated on their *front* surfaces. Light does not pass through them and so is not subject to refraction and color dispersion as in a prism or a lens. Telescope (and photographic camera) lenses can compensate for this effect, but only over a relatively small wavelength range.

In spite of the many advantages of reflecting telescopes, refractors remained the preferred instruments of astronomers for centuries. They were more luminous because the early reflecting telescopes' mirrors were metallic and had poor reflectivity. They are also less sensitive to optical misalignment. Thanks to technological advances, however, reflecting telescopes finally supplanted them late in the nineteenth century.

203 What does a large modern telescope look like?

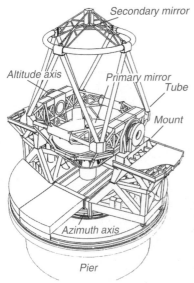

Drawing of ESO's VLT.

The figure on the left represents a large modern optical telescope. The main optical elements – the primary and secondary mirrors – are located inside what is traditionally called the "tube." Originally, the tube of a telescope really was a wooden or metal tube that enclosed the incoming optical beam, with the primary mirror located at the bottom of the tube and the secondary mirror at the top. Although the term "tube" is still used, the supporting structure has now usually been reduced to a simple truss in order to save weight and help it match the temperature of the ambient air.

The tube is supported by the *mount* which itself is supported by a *pier*, often made of concrete and well anchored in the ground to ensure stability. The pier should be high enough (10 m or more) to place the telescope above the layer of air closest to the ground, which is frequently subject to turbulence that blurs the image.

The telescope can be rotated around two axes in order to point it towards a celestial object and follow it as it moves across the sky from east to west, due to the Earth's rotation. These movements must be extremely smooth so as to avoid vibrations that would degrade the image. To accomplish this, the telescope is generally mounted on "hydrostatic pads" into which oil is injected under pressure to lift the instrument slightly. The telescope then glides on a thin film of oil, rotating smoothly and with almost no friction.

Nearly all telescopes built before 1980 used the equatorial mount, which provides for motion both around the "polar axis," parallel to Earth's rotation axis, and around the "declination axis," perpendicular to it. Once pointed toward a target, the telescope can be kept on target by rotating it around the polar axis at the Earth's rotation rate, but in the opposite direction. This straightforward drive system was a decisive advantage in the early days: a simple clock mechanism was all that was needed.

Equatorial mounts are intrinsically heavy because of the inclined polar axis and, in practice, can only be used for telescopes up to 5 m in diameter. Larger telescopes must be supported more efficiently by placing the center of gravity of the tube above the base of the mount. This is the principle behind the altitude–azimuth mount.

In the altitude–azimuth mount, "alt–az" for short, the tube is oriented by rotation around a vertical axis (the azimuth axis) and a horizontal axis (the altitude axis). Unlike equatorial mounts, neither of the two axes supporting an alt–az mount changes direction with respect to gravity. Structurally, it is the sturdiest and simplest mount. The reduction in mass (and cost) is so significant that it has now become the standard mount even for mid-class telescopes.

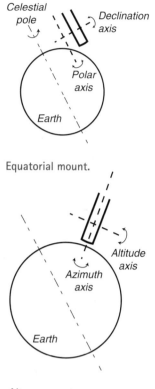

Equatorial mount.

Alt–az mount.

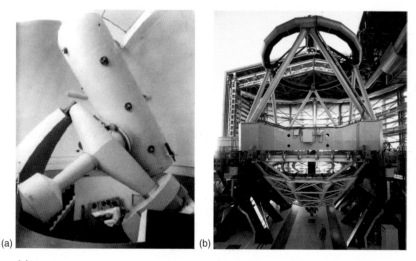

(a) A telescope with an equatorial mount (Haute Provence Observatory). Note the counterweight balancing the weight of the tube. (b) An example of an alt–az mount (ESO VLT).

Three axes of rotation are needed, however, not just two as with equatorial mounts. Two are required to orient the tube (alt and az) and a third to compensate for "field rotation," as the field of view rotates around itself while the telescope tracks an object. Also, during tracking, each of these three axes must be rotated at *variable* speeds. With the advent of computer control, however, this complication is easily handled.

204 What are the most common optical configurations?

In an optical telescope, light from a celestial source is collected by a concave mirror, called the *primary mirror*, which concentrates the light at its focus. The light can be analyzed there directly or, alternatively, the beam can be sent on to another focus that is either more accessible, has a larger scale ("magnification"), or is more convenient to work with in conjunction with scientific instruments such as cameras and spectrographs.

Most modern telescopes are designed to be used at a number of different foci. It is a little like using a reflex camera, where the choice of lens depends on the scene to be photographed. The most common configurations are shown hereafter.

The **primary focus** has the smallest number of reflecting surfaces possible, important for maximizing optical efficiency.[†] The focal ratio[‡] is usually between $f/1.5$ and $f/3$ so as to minimize the length of the tube and the diameter of the dome. This range provides a good focal scale for wide field observations, but is too small for most other uses. One variation of the primary focus is the Newtonian focus, in which the prime focus is relocated outside the incident beam by means of a tilted flat mirror. Although often used on amateur telescopes, this configuration is seldom found on telescopes with

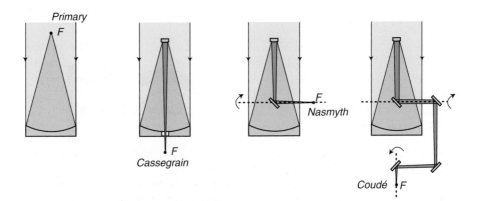

Configurations of the most common foci.

[†] Between about 3 to 15% of light is lost for every mirror in the train, depending on the type and condition of the reflecting surfaces and the amount of dust on them.

[‡] The focal ratio, or f-number, of a telescope focus is the ratio of its focal length to the diameter of the primary mirror. For example, if the primary mirror is 10 m in diameter and the focal length is 20 m, the focal ratio of that focus is $f/2$.

very large mirrors because there the partial obstruction by the analyzing instrument at the prime focus is not a problem.

In the **Cassegrain focus** a convex mirror sends the beam behind the primary mirror which has an opening at the center. This is the most commonly used focus and some telescopes have no other. The focus is easily accessible, the optical efficiency is good because there are only two reflecting surfaces, and the field covered can be relatively large, since the use of two optical surfaces permits excellent correction. The aperture is generally between $f/8$ and $f/16$, providing a focal scale that works well with most instruments. The location of this focus near the primary mirror – and thus near the tube's center of gravity – means that relatively heavy instruments can be used with it.

The **Nasmyth focus** is a Cassegrain whose beam has been folded back along the tube's axis of rotation. It is often used in alt–az telescopes since analyzing instruments can be installed on the mount, making them easily accessible.

The **coudé focus** is a Cassegrain whose long focus is sent back through the telescope's axes of rotation in such a way as to remain fixed in space when the telescope swivels. Useful with large, heavy instruments such as high-resolution spectrographs, the coudé focal aperture is set between $f/30$ and $f/100$ so that the diameter of the beam, and thus of the flat reflecting mirrors, will be small enough. The number of flat mirrors depends on the type of mount, but usually falls between 5 and 7. The main drawback of the coudé focus is thus poor optical efficiency, especially at short wavelengths where mirror reflectivity is poor. Optical fibers fed by the prime or Cassegrain focus are used more often today.

205 How is the performance of a telescope measured?

It is useful to have a simple way of describing an optical instrument to characterize its performance. For a traditional camera, this would be the film size (35 mm or 6×6 cm), for a digital camera, the number of pixels and the focal length of the lens, and for a microscope, the maximum magnification. What about a telescope?

The most important characteristic of an astronomical telescope is the *diameter* of its primary mirror. This determines:

- its *sensitivity*, meaning its ability to observe faint objects: the larger the diameter, the larger the surface available to collect the faint incoming light;
- its *resolution*, for the larger its diameter compared to the wavelength of incoming light, the less the image will be smeared by diffraction (Q. 211). However, that is only true for space telescopes or for ground telescopes that are corrected for atmospheric turbulence (Q. 219). Telescopes that are affected by the atmosphere all have about the same limit to resolution, on the order of one second of arc, regardless of their diameters (Q. 213).

Other factors such as the amount of sky coverage, the plate scale (magnification) of the foci, the ability to support large instruments, etc., are important scientifically, but those factors are secondary or modifiable.

206 What is the shape of a telescope mirror?

A concave mirror reflects light and concentrates it, but the image it produces is not necessarily good. In the simplest case, a spherical mirror, the image of a point source at infinity, such as a star, is not a point but a small disk – the image is said to be degraded by *spherical aberration*.

Among all concave surfaces, only one shape produces a perfect image of a point at infinity, the paraboloid, which is the surface generated by a parabola rotated on its axis. Let us understand why.

When light strikes a flat mirror, it bounces off symmetrically with respect to a perpendicular to the surface (Q. 210), and a curved mirror can be considered as being made up of a series of infinitely small flat mirrors. The parabola has the following wonderful property: the normal (i.e. the line perpendicular to the tangent) at any point on its surface bisects the angle formed by the straight line parallel to its axis and the line connecting that point to what, in mathematics, is called its focus. This exactly corresponds to the law of reflection. This being true at all points on the parabola, the light rays coming from an object at infinity on the axis of a parabolic mirror will all, after reflection, pass through the parabola's focus.

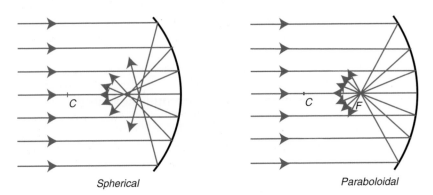

Spherical   Paraboloidal

A spherical mirror does not produce a good image of a point source at infinity because the reflected rays do not all pass through the same point. But they do in a parabolic mirror.

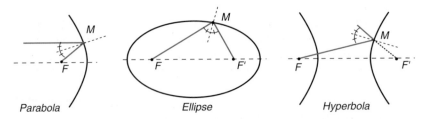

Parabola   Ellipse   Hyperbola

The property of conics that is the foundation of reflecting telescopes: the normal at any point of a conic bisects the two radii issuing from that point (Apollonius's theorem).

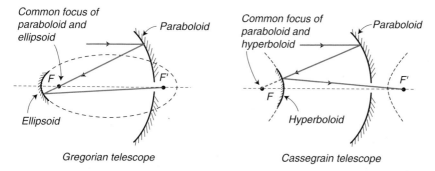

Gregorian telescope   *Cassegrain telescope*

A system of coaxial conic surfaces with coincident foci is "stigmatic."

This property of the parabola is common to all conical surfaces: paraboloids, ellipsoids, and hyperboloids. The normal at any point bisects the angle formed by the two radii joining that point to the two foci, a property described by the Apollonius theorem.[†] This simply means that all optical rays issuing from a source located at one of the foci will converge at the other focus, forming a perfect image of the source. This perfect imaging of a point source is called "stigmatism."

This property of conics is at the heart of the optics of reflecting telescopes. The simplest case is the parabola, which is a degenerate ellipse with one focus at infinity. Rays issuing from this focus at infinity (i.e. rays parallel to the parabola axis) will, after reflection, converge at the parabola focus.

The next case uses a second conic surface with one of its foci coincident with the focus of the parabola. This second conic surface will re-image the original source at its second focus, again in a perfectly stigmatic way. If this second conic is an ellipsoid, the system is called a "Gregorian," and if it is a hyperboloid, it is a "Cassegrain." A system with three powered mirrors follows the same principle. The first and second mirrors of a two-mirror system are called the "primary mirror" and the "secondary mirror," respectively. The third powered mirror of a three-mirror system is called the "tertiary," and so on.

207 How are telescope mirrors made?

The key piece of any telescope is its primary mirror. Its diameter determines the telescope's collecting power and the quality of its surface determines the quality of the images formed.

The primary mirror is also the most demanding piece to fabricate because the optical surface must be shaped to within a fraction of a wavelength (about twenty thousandths of a millimeter in the visible), and must not deform when temperatures change or when the telescope is rotated. As a comparison, if an area the size of the US was flattened to

[†] Apollonius (262–190 BC), one of the greatest mathematicians of antiquity, was the first to understand that a spherical mirror could not furnish a perfect image, as he explained in his treatise *On the Burning Mirror*. He coined the terms parabola, hyperbola, and ellipse, and demonstrated many of their properties including this famous theorem.

the same precision as a 10 m telescope mirror, no pimple or pit over 2 cm would mar the surface!

To reach that level of perfection, there are three conditions:

1 The material used for the mirror must take a very high polish.
2 The material must expand and contract very little with changes in temperature.
3 The mirror must be supported to prevent its sagging under the effects of gravity.

As far as materials are concerned, nothing is better than glass or its derivatives (vitrified ceramic or silica).[†] Glass takes an excellent polish, better than metals. It can also be treated so that its coefficient of expansion is almost zero, much lower even than that of Pyrex.

As for the effects of gravity, preventing a mirror several meters in diameter and weighing several tons from sagging more than half a ten thousandth of a millimeter is no small matter. The solution consists of supporting its back and rim with systems that compensate for the effects of gravity no matter how it is oriented. The mirror is then "weightless" for all intents and purposes.

Polishing techniques have changed little over the centuries. The surface of the mirror blank is first covered with a slurry of water and abrasive powder, then ground down by rubbing it against a second block of material called the "tool." Since mirrors are usually symmetric about their axes, the mirror blank is turned on its axis while the tool is moved back and forth sideways. The mirror blank and the tool grind each other down mutually, tending to assume a spherical shape because a sphere is the only surface that fits perfectly against itself when rotated or moved in any direction.

The three phases of polishing include:

• grinding, where a coarse abrasive and a tool with the same diameter as the mirror blank are used to rapidly create a spherical surface approximating the one to be finished;
• polishing, where finer and finer abrasive powders are employed to reduce defects on the mirror's surface and make it reflective enough for optical testing;
• figuring, where the mirror is given its final shape, usually parabolic or hyperbolic.

For a large mirror, this last phase, the longest and most exacting one, can require a year or more. Since the blank's surface is no longer spherical, a rigid, large-diameter tool cannot match its shape over the entire surface. Modern techniques consist of

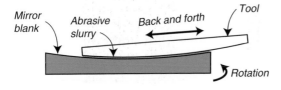

Principle of mirror polishing: the mirror blank is rotated around its axis while the tool rubs back and forth against it.

[†] Other materials such as beryllium or silicon carbide are occasionally used for the sake of lightness, but mainly for space applications.

Polishing an 8 m mirror using a small computer-controlled tool. The optician in the background gives an idea of the scale. Credit: REOSC.

either deforming the tool as it is moved or using a small tool and regulating its pressure and position very precisely. In either case the entire process is computer controlled.

208 What is a Schmidt telescope?

This type of telescope was invented in 1930 by the Estonian-German optician Bernhard Schmidt for wide field astrophotography. Its principle is quite elegant, consisting of using (a) a *spherical* rather than parabolic mirror so as to eliminate off-axis aberrations

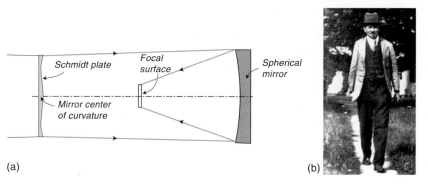

(a) Schematic of the Schmidt telescope (the aspherical shape of the plate is exaggerated for visibility). (b) Bernhard Schmidt (1879–1935), who was a master glass polisher despite having lost his right hand at the age of 15 while experimenting with explosives.

(since a spherical mirror is symmetrical in all directions[†]) and (b) an *aspherical* glass plate to correct the strong aberration caused by the spherical mirror.[‡] The Schmidt telescope produces excellent images of fields covering several degrees. Its drawbacks are that the focal surface is not flat (making it necessary to bend the photographic plates) and that the tube is nearly three times as long as that of a classic, two-mirror telescope. The glass plate is a complex shape and difficult to fabricate, but Schmidt conceived the idea of deforming it by applying suction on its lower surface and polishing it flat. When the suction is released, the plate assumes the desired shape.

Very few Schmidt telescopes have ever been constructed, but those few have been extremely useful for making all-sky surveys and for detecting comets and asteroids.

209 Why are telescopes housed in domes?

Unlike radio telescopes which sit out in the open, fragile optical telescopes need protection from harsh weather. But why use a dome shape? First, the enclosure has to be circular because it must rotate on rails to keep its opening (slit) facing the telescope no matter where it points. Second, using a spherical shape is the simplest way to allow the telescope and the enclosure to move independently without risk of collision. Some observatories have cylindrical enclosures, however.

Domes are complex, heavy structures up to 30 m in diameter that must rotate smoothly to avoid transmitting vibrations to the telescope during observations. They also play a role in the performance of the telescope that they shelter. If the telescope and ambient air are not at exactly the same temperature during observations, image quality will be degraded by convective air movements. Hence, during the day, the telescope cannot be allowed to heat up; the dome must be well insulated and have a highly

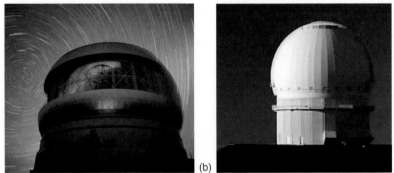

(a)   (b)

Two examples of domes. (a) The dome of the Gemini South Observatory with its ventilation louvers open. (b) The Canada-France-Hawaii Observatory in Hawaii. Credit: Gemini Obs. and Cuillandre.

[†] A classic parabolic mirror is only symmetrical for rays that arrive parallel to its axis. On-axis images are perfect, but off axis they are marked by aberrations (coma and astigmatism) of horrendous proportions when fields are even slightly wide.

[‡] Such optical systems composed of both mirrors and lenses are said to be "catadioptric."

reflective outer skin. Air conditioning and cooling floors can also help cool its interior. As night falls, the thermal exchange between the telescope and ambient air must be maximized so as to eliminate differences in temperature as quickly as possible, then allow the telescope to match temperature variations throughout the night. This calls for large louvers to ventilate the dome by natural means or even for active ventilation.

Although observatories are strictly functional structures in which architectural aesthetics plays almost no part, domes often evoke strong feelings. With their pure forms perched between Earth and sky on lofty mountaintops, they are like modern cathedrals dedicated to unraveling the mysteries of the Universe.

210 Reflection, refraction, diffusion, dispersion ... want a short refresher?

When a ray of light strikes a smooth, transparent surface, part of it bounces off symmetrically with respect to a perpendicular to the surface. This phenomenon is called *reflection*. The other part of the ray penetrates but changes direction because the material causes the velocity of light to slow down. The deflection is a function of the change in speed, which is quantified by the material's refractive index, n. This is the phenomenon of *refraction*. A glass surface reflects about 4% of the light, while the rest of it penetrates and refracts. If the surface were metallic, up to 99% would be reflected, the remaining being absorbed.

If we examine the refracted ray more closely, we notice that its light has been decomposed into different wavelengths, for a material's refractive index varies with wavelength. This is the phenomenon of *dispersion*.

If the surface is not smooth (compared to the wavelength of the light) the light rebounds, but in any and all directions. This is the phenomenon of *diffusion*. It also occurs when light strikes small particles, such as dust, either suspended in the air or out in space.

The phenomenon of *diffraction* calls for a somewhat lengthier explanation, which can be found in question Q. 211.

211 ... and diffraction?

We are so used to seeing light travel in a straight line that it is hard to imagine it *also* traveling "sideways," but that does happen. This is because light can behave like a wave, and a wave spreads out when its advance is partially blocked. Surface waves

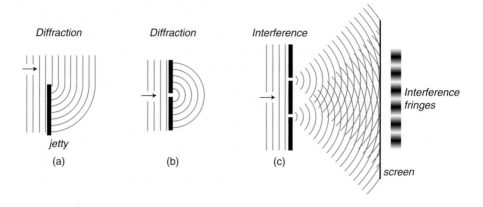

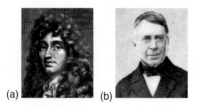

(a) Christian Huygens (1629–95), physicist and astronomer, one of the greatest scientists of all times. He was the first to understand that light can behave like a wave. (b) George Airy (1801–92).

in water show this effect nicely. When waves arrive at a jetty in a harbor, they bend around it and will even be found, albeit weakened, directly behind it (as in *a* below). If there is a small hole in the jetty (as in *b*), they emerge from it in the form of a set of circular waves centered on the opening. Such a geometrical dispersion of light rays is the phenomenon called *diffraction*.

If there are two holes or slits, the two wavefronts emerging from them "interfere" with each other, the resulting wave being the sum of the two wavefronts: a trough in one of the wavefronts and the crest from the other cancel each other, two troughs combine to form a deeper trough, two crests combine to make a higher crest. This is the phenomenon of *interference*. If, instead of water, we are dealing with light, and if we collect on a screen the light which has passed through two slits, we would see a pattern of successive light and dark bands called *interference fringes*.

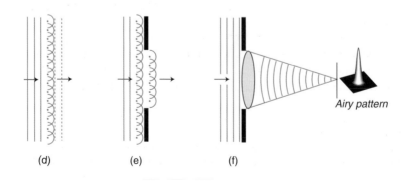

To return to diffraction, the circular shape of the wavefront emerging from a slit (as in (b)) led the Dutch physicist Christian Huygens to suggest that a light wave travels as though each point along the wave was the source of a small new train of waves, a "wavelet," the wavefront being the result of the superposition of all these wavelets (as in (d) in the next figure). So if the light from a star comes to the aperture of a telescope, only some of those wavelets will enter it (e), and the image formed by the telescope will be the result of the interference of those captured wavelets (f). An infinitely large aperture would capture all the "information" from the source: the interference of the wavelets would be total and perfect, the image of a point source would be a point. With a partial capture, however, the image is not "complete"; instead of being an infinitely small point, the image is a small, bright "spot." That spot is called the *Airy disk*, from the name of the English astronomer who first described it mathematically.

Close analysis of the spot shows that it is actually composed of a bright disk surrounded by alternating bright and dark concentric rings which become progressively fainter with distance from the center, a phenomenon called the *Airy pattern*.[†]

The Airy pattern extends out to infinity, but most of the light is concentrated in the central disk and the nearest bright rings: 84% of the light is actually inside the first dark ring. By convention, the size of the Airy disk is defined as the diameter of that first dark ring, which is given by $2.44\lambda/D$, where D is the diameter of the telescope and λ is the wavelength of the observed point source. As one might expect, the larger the telescope, the smaller the size of the Airy disk.

No telescope can produce an image of a point source (a star, for example[‡]) smaller than the Airy disk, and even then, the optics must be perfect and unaffected by atmospheric turbulence. Space telescopes such as the Hubble telescope can accomplish this. For ground telescopes, the size of the image is larger because of atmospheric turbulence unless compensated by adaptive optics (Q. 219).

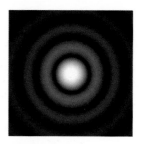

The Airy pattern.

212 How is the resolving power of a telescope defined?

The angular or spatial resolution of an optical system is the smallest detail that can be distinguished in the image that it forms of an object. In the particular case of astronomical images, angular resolution is defined as the smallest angle between two distinguishable point sources.

Resolution is obviously affected by imperfections in the optics and, in the case of ground-based observations, by atmospheric turbulence. But for a perfect optical system unaffected by the atmosphere, resolution is limited by the size of the Airy disk (Q. 211). In such a case, since the size of the Airy disk is inversely proportional to the diameter

---

[†] The formula for the distribution of brightness intensity in an Airy pattern is rather complex and involves Bessel functions.

[‡] Stars are point sources for all practical purposes. Even the most powerful telescopes cannot make out their disks – for that, a special instrument called an interferometer is needed (Q. 227).

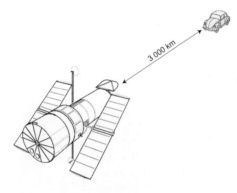

When two stars are close together, their combined images resemble that of a single star (a). As their separation increases, the peak of their combined image flattens out (b), then dips in the center (c).

of the telescope, the larger the telescope, the better the resolution. The British physicist Rayleigh proposed that two images should be considered resolved when the peak of one falls on the first ring of the other (c). This condition, which is largely accepted, is called the *Rayleigh criterion*. For perfect optics unaffected by atmospheric turbulence, this criterion corresponds to an angular separation of $1.22\,\lambda/D$, where λ is the wavelength and D the diameter.

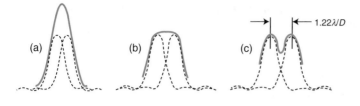

The Hubble Space Telescope, with a primary mirror 2.40 m in diameter, has an angular resolution of 0.06 arcsecond. That is about 10 times better than ground telescopes can do, since their resolution is limited by the atmosphere regardless of their diameter. That gain in resolution explains Hubble's fantastic scientific harvest. Just how extraordinary is its resolving power? Well, if the Earth was flat and without an atmosphere and the Hubble telescope was sitting in New York City, it could distinguish the two headlights of an automobile ... in Mexico City!

213 Do celestial objects look bigger through a large telescope?

This seems an obvious question to ask, but how much a telescope magnifies an image is not one of its fundamental characteristics. If we are observing visually, simply choosing an eyepiece with stronger magnification makes the image look "bigger." And if we are taking photographs, we can easily enlarge them afterwards. What is really important is not how big the image is but the amount of detail it provides. In other words, the telescope's *resolution* (Q. 212).

In theory, the greater a telescope's diameter, the better its resolution. Unfortunately, even with perfect optics, images made by ground telescopes are degraded by atmospheric turbulence (Q. 218). Increasing the diameter of a telescope over and above about 15 cm (6 in) is fruitless: the resolution of a large professional telescope is not much better than that of an amateur telescope. Improvement can only be had by choosing a site with less atmospheric turbulence.[†] That is why the great modern observatories are

---

[†] Adaptive optics can compensate for some atmospheric turbulence (Q. 219) but work best if seeing is already good (Q. 218).

mostly concentrated in a limited number of sites (Q. 221). It is also why observations made from balloons and from space are so prized.

214 Who invented the telescope?

Strangely enough, no one really knows. The British Museum in London holds the oldest known lens, which was found in the ancient Assyrian city of Nimrud (in present-day Iraq). It may have been used as a magnifying glass, but is not of sufficient quality to have been part of a telescope. Correcting lenses for near- and far-sightedness were known in ancient times – the Roman emperor Nero watched gladiator fights through a monocle – and they were fairly common in Europe as early as the thirteenth century. But it was not until the early 1600s that Dutch lensmakers discovered that a converging lens placed in front of a diverging one made objects appear closer. By 1608 Hans Lippershey, Zachery Janssen, and others were turning out telescopes in series called *Dutch perspective glasses* for military use. Their quality was mediocre, and no sky observations are known to have been made with them.

Fresco showing a medieval churchman wearing glasses – Treviso, Italy 1352.

Galileo heard about this invention and made his own telescope in 1609, proposing it to the Doges of Venice to help identify distant ships. This first telescope of his had an angular magnification of only three, but he quickly perfected it and, in 1610, constructed one with a magnification of 33. It was with this instrument that he discovered the satellites of Jupiter, sunspots, the phases of Venus, and the mountains and valleys of the Moon (Q. 194). The original inventor of the telescope, completely overshadowed by Galileo's technical improvements and brilliant observations, was forgotten, and this early type of telescope is now called a *Galilean telescope*.

In the Galilean telescope, a divergent lens is used for the eyepiece. Then, in 1611 Kepler showed that a convergent one could be used just as well, and this is the configuration that became standard for astronomical telescopes, its advantages being that it can have a much higher magnification, a wider field, and that a micrometer can be placed at the focus of the eyepiece to make angular measurements. The image it produces is upside down, which would be a problem for terrestrial use but is of no consequence in astronomy. A few decades later Huygens built several telescopes of this type with which he went on to discover Titan, Saturn's brightest satellite, in 1655.

Galileo (1564–1642) and two of his telescopes, now preserved in a Florence Museum.

215 What major improvements have been made in telescopes since Galileo's time?

The main defect of early telescopes was chromatism (Q. 202). Isaac Newton, who in 1666 discovered the dispersion of light by prisms, was the first to understand that it was due to the lenses, and decided to use mirrors instead. In 1671 he successfully constructed a telescope with a spherical bronze mirror 16 cm in diameter. That idea was not pursued, however, because metallic mirrors of that time had mediocre reflectivity and were difficult to polish into parabolic shapes.

Herschel's 20 ft telescope.

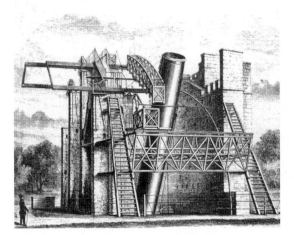

Lord Rosse's Leviathan.

Little changed until 1758, when John Dollond invented the achromatic lens, a compound lens composed of two juxtaposed lenses, one convergent, one divergent, made of glass with different indices of refraction. This advance in quality was an important advantage for astrometry, the precise measurement of stellar positions.

William Herschel, whose ambition it was to determine the shape and size of the Universe, needed greater sensitivity and better angular resolution than could be obtained with the refracting telescopes of his time. In other words, he needed a telescope with a larger diameter. Unfortunately, glassworking techniques back then could produce good quality lenses no larger than about 10 cm in diameter, so Herschel turned back to metal mirrors. In 1783, after conducting numerous experiments, he produced the most powerful instrument of his day, a 20 ft-long telescope (6 m) with a bronze mirror 50 cm in diameter, with which he made his most important observations. Later, he built one 40 ft (12 m) long with a 120 cm mirror, but it had so many flaws that the 20 ft instrument continued to be his favorite.

The Herschel telescopes remained unsurpassed for more than half a century. Although as brilliant a telescope maker as he was an observer, Herschel left behind no notes about his fabrication methods. When the rich Irish amateur astronomer William Parsons (Lord Rosse) wanted to build an even bigger telescope, he had to reinvent everything, from the exact composition of the bronze alloy to polishing techniques. But he eventually managed to build a telescope 6 ft (180 cm) in diameter. Completed in 1845, it was by far the largest of its time. As a matter of fact, its nickname was "The Leviathan," and Jules Verne spoke admiringly of it in his prophetic novel "From Earth to the Moon." Even though the telescope's performance was severely criticized ("Lord Rosse's telescope is a joke!," sniped Léon Foucault), its optical quality was actually quite decent, although degraded by atmospheric turbulence at its unquestionably mediocre site. In any case, it was by far the most sensitive telescope of its day, reaching magnitude 18. Rosse and the professional astronomers he invited made remarkable observations with it, resolving individual stars in certain galaxies for the first time and discovering more than a dozen spiral nebulae. In operation for 30 years, it was only surpassed in size in 1917, when the 100 in (2.5 m) telescope was built on Mount Wilson.

Meanwhile, early in the nineteenth century, Pierre Louis Guinand, a Swiss optician and associate of the German physicist Joseph von Fraunhofer, invented ways of casting and polishing lenses of greater diameter than before. That gave new impetus to refracting telescopes, which many preferred to reflectors because their metal mirrors had to be repolished every few months. As a result, observatories equipped themselves with refracting telescopes with ever larger diameters, culminating with the giants of the late nineteenth century, including the 80 cm Paris-Meudon and 40 in (1 m) Yerkes refractors.

Observers could not simply rely on obtaining ever larger lenses, however, because lenses over a meter in diameter sag seriously under their own weight. In addition, it is difficult to obtain large glass blanks of sufficient purity.

Hence, when larger-diameter instruments were needed for the observation of even fainter objects, mirrors returned to favor. The crucial advantage of a mirror is that it can be supported from behind without affecting the optical beam. Moreover, it need not be of optical-quality material because light does not go through it. All it has to do is take a good surface polish. And for that, glass is better than the bronze that was being used. Glass is not very reflective, but a technique for depositing a fine, smooth

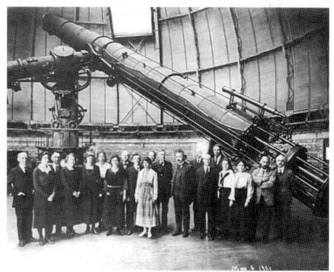

The 40 in (1 m) Yerkes telescope at the University of Chicago, the largest refractor ever built, here shown in 1921 during a visit by Albert Einstein (right of center).

Léon Foucault (1819–1868) and his 80 cm telescope.

layer of silver on a mirror's *front* surface[†] had just been invented, and Léon Foucault applied this technique to astronomical mirrors. He had several telescopes with silvered mirrors built, including, in 1862, the Marseille Observatory 80 cm instrument, which was operational for almost a century.

Today, coating under vacuum with aluminum, gold, or protected silver has replaced the chemical silvering technique once used, but the concept of the reflecting, glass mirror telescope developed by Foucault continues to serve as the basis of essentially all optical telescopes built during the past 150 years. The modern astronomical telescope had been invented.

216 Why do astronomers want ever-larger telescopes?

Scientific knowledge is not bound by limits. A given science can slow down, of course, even almost exhaust its subject: classical mechanics and human anatomy may require small adjustments or more detailed studies here and there, but not much remains to be discovered. In the late nineteenth century, when our knowledge of the Universe was limited to our own galaxy, some felt that the practical end of astronomy was also near. Subsequent advances in physics and technology have proved them wrong. As in the case of most other sciences, each new advance has raised new questions or opened new fields of study.

Progress in understanding the Universe always requires new data, and for that we need to "see" better, further, and differently. To see better, telescopes need higher angular resolution. To see further, they must collect more photons. To see differently, they must explore new domains of information in the cosmos, for example by probing other regions of the electromagnetic spectrum or by analyzing the radiation of new particles, such as neutrinos.

To see better and further, the only solution is to increase the diameter of telescopes, since angular resolution improves as the diameter of the primary mirror (Q. 212)[‡] and the collecting power increases in proportion to the area of the primary mirror, that is, as the square of its diameter.

To appreciate the importance of a telescope's size when observing deep sky objects, think of this: the most distant galaxies (those with a magnitude of 30 in the visible) are

[†] Household mirrors have a deposit of tin-silver alloy on the *back* surface of the glass. This guarantees a smooth reflective surface and protects from oxidation, but back-coating cannot be used for astronomical mirrors, since they need to reflect from the *front* surface to avoid chromatism.

[‡] This only became true about 30 years ago, when it became possible to eliminate the effect of atmospheric turbulence by going to space or by using adaptive optics.

a thousand billion (10^{12}) times less luminous than Sirius, the brightest star in the sky. In the visible domain, we receive from those distant galaxies less than 1 measly photon per second and per square meter!

This need for sensitivity and resolution is the reason why there has been a systematic increase in telescope size ever since Galileo. Historically, the largest telescopes have tended to double in size every 40 years [35]. This exponential increase might be considered the equivalent of Moore's law in electronics, which states that the number of transistors on a printed circuit doubles every two years.

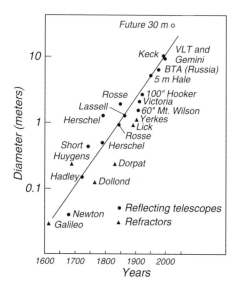

217 What are the largest optical telescopes today?

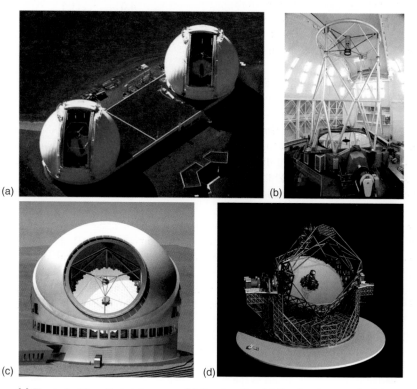

(a) The twin 10 m Keck telescopes; (b) the 8 m Gemini North telescope, both in Hawaii. Artist's views of (c) the future TMT, and (d) E–ELT.

The largest optical telescopes.

Telescope	Country	Location	Diam. (m)	Year completed
LBT[1]	USA	Arizona	11.8	2008
Keck I and II	USA	Hawaii	10.5	1993/1998
GTC	Spain[2]	Canary Islands	10.4	2008
Hobby–Eberly[3]	USA	Texas	9.5	1999
SALT	South Africa[2]	South Africa	9.5	2007
Subaru	Japan	Hawaii	8.4	1999
Gemini N and S	International[2]	Hawaii/Chile	8.3	2000/2002
VLT (4 telesc.)	Europe	Chile	8.2	1998 to 2001
MMT[4]	USA	Arizona	6.5	1999
Magellan I and II	USA	Chile	6.5	1999/2002
BTA	Russia	Caucasia	6.0	1976
LZT[5]	Canada	Canada	6.0	2004
Hale	USA	California	5.1	1949
WHT	UK	Canary Islands	4.2	1987
SOAR	USA[2]	Chile	4.2	2004
Blanco	USA	Chile	4.0	1976
Lamost	China	China	4.0	2004
AAT	Australia	Australia	3.9	1975
Vista	UK	Chile	3.9	2008
Mayall	USA	Arizona	3.8	1973
UKIRT	UK	Hawaii	3.8	1978
CFHT	Canada/France	Hawaii	3.6	1979
ESO 3.6	Europe	Chile	3.6	1976

1. Two 8.4 m primary mirrors on same mount.
2. These observatories are international or have international partners.
3. Fixed altitude mount; equivalent diameter.
4. Originally equipped with six 1.8 m mirrors
   (equivalent to a single 4.5 m mirror) and commissioned in 1979.
5. Liquid mirror used in transit.

The USA dominated observational astronomy during the first three quarters of the twentieth century, the largest optical telescopes all being on American soil. First came the famous 60 in (1.50 m) and 100 in (2.50 m) telescopes on Mt. Wilson near Los Angeles, and later a 200 in (5 m) on Mt. Palomar in southern California, completed in 1949. This giant was surpassed in size only in 1976 by a 6 m Soviet Union telescope located in the Caucasus. Hindered by poor seeing and mediocre optics, however, it was not very productive scientifically. Still, as the first modern optical telescope with an alt–az mount, it was an important milestone.

America began to face competition in the 1970s, when six new large telescopes were built, two by the USA but also one each by Australia, the UK, Europe (ESO), and France/

Canada. All these were slightly smaller (3.6 to 4 m) than their predecessors, but were just as efficient because they benefitted from good astronomical sites and new technology.

Electronic detectors, which appeared in the 1980s, were so much more efficient than the photographic plate used until then that, for a while, astronomers' incessant demands for greater sensitivity could be met without resorting to larger collecting surfaces. Eventually, though, these became necessary and led to the construction of several 8–10 m diameter telescopes in the 1990s.

In practice, casting difficulties and road transport limit the size of single mirrors to 8 m. Beyond that the solution is to "segment" the primary mirror, assembling it from small adjacent mirrors in a honeycomb configuration. The difficulty there is that the individual mirrors must be positioned with extreme accuracy with respect to each other (on the order of a fraction of the wavelength, i.e. about one ten-thousandth of a millimeter) so that they mimic a solid, one-piece mirror. This solution was initiated with great success by the 10 m Keck telescopes.

Telescopes of even larger size are currently under study, all with segmented mirrors, including the TMT (*Thirty-Meter Telescope*) in the USA and the 42 m E-ELT (*European Extremely Large Telescope*) in Europe.

218 How does the atmosphere degrade telescope images?

The atmosphere is never at rest; updrafts of air heated by the ground and local and global winds keep it in constant motion, creating the "atmospheric turbulence" familiar to airplane travelers. This turbulence is also responsible for degrading our telescopic images of celestial objects. The index of refraction of air is hardly affected by the pressure variations that accompany air movements and can shake an entire airplane, but it is unfortunately extremely sensitive to changes in air temperature.

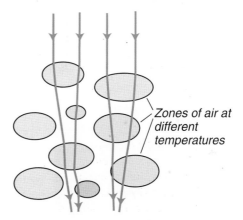

Zones of air at different temperatures

It is this same phenomenon that causes the wavy visual distortions above a road on a hot day. Just as a ray of light is bent when passing from air to water or glass, it is bent when passing from a region of slightly warmer air to a region of slightly cooler air, or vice versa. Rays of starlight that are nicely parallel when they reach the outer edges of our atmosphere are no longer parallel when they arrive at a telescope. Instead of converging at a single point at its optical focus, they spread out a bit, smearing and blurring the image. And since air masses are always in motion, the image moves continuously while its intensity varies, a phenomenon called "seeing" in the jargon of astronomers. This is why stars "twinkle." Small telescopes (amateur telescopes in particular) are strongly affected by scintillation (changes of intensity) and image motion. Large telescopes, which collect the light over a greater area, "average" these effects and their images are almost stable in intensity and position, but still they are smeared and blurred. Regrettably, this effect makes the image 10 to 100 times larger than such telescopes would otherwise be capable of providing (Q. 211). Their theoretical resolution is thus never reached, unless the new technique of adaptive optics is used (Q. 219).

219 What is adaptive optics?

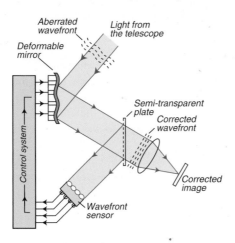

Basic principle of adaptive correction of atmospheric turbulence. The wavefront error is analyzed, then corrected with a deformable mirror.

Traditional optical systems are composed of optical elements (lenses and mirrors) which are rigid and firmly positioned with respect to each other. This is to minimize image degradation caused by the deformation of the optical surfaces and misalignments when changing orientation. Traditional optics is solid as a rock. Adaptive optics, on the other hand, is, as its name implies, composed of *movable*, *flexible* elements. The intent is to make real-time corrections for perturbations that may affect image quality. It works on the principle of the servo system. Just as the cruise control on your car maintains a constant speed regardless of the grade of the road, an adaptive optics system keeps an image as stable and sharp as possible by analyzing it and immediately compensating for whatever affects it. Adaptive optics has many applications, but in astronomy its main purpose is to correct for atmospheric turbulence (Q. 218).

The method consists of observing a bright star close to the celestial field under study and determining how its light rays have been affected by their passage through the atmosphere – that is, by measuring what is technically called the *wavefront error*. This information is then used to deform a flexible mirror in such a way as to introduce a wavefront error of opposite sign, to compensate exactly for the initial error. The light beam emerging from the system is then as if it had never been perturbed. The image of the bright reference star is almost perfect, and the image of the celestial object of interest

Laser beam creating an artificial star in the atmosphere. Credit: Gemini Obs.

The center of the Galaxy in infrared (false colors) without and with adaptive optics. Credit: UCLA/Keck.

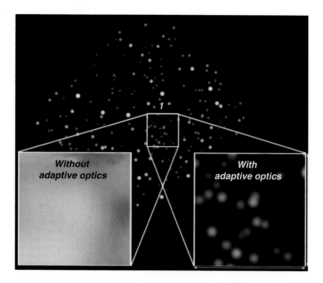

is, also – provided that it is sufficiently close angularly that atmospheric turbulence affects its rays in the same way.

This technique has been used very successfully, but is unfortunately applicable to only a small fraction of the sky. This is because, when measuring a wavefront error, the reference star must be bright enough for the error to be measured on a timescale compatible with the turbulence, i.e. about a thousand times per second. Unfortunately, there are not many stars that bright in the sky.

To cover the entire sky, the solution is to create an "artificial star" high in the atmosphere near the object of interest. This is typically done by focusing a powerful laser beam on the "reflective" layer of sodium found in the atmosphere at an altitude of about 95 km. The equipment needed, although complex and costly, has been recently installed on a few large telescopes and allows them to recover a good part of their theoretical resolution and greatly increase their sensitivity. The technique has its limitations: it is currently only applicable in the infrared where atmospheric turbulence is less severe than in the visible, and the field of good compensation is relatively small. Even so, the scientific results have been very promising.

220 Are there any alternatives to traditional mirrors?

To form an image of a distant source, its light rays must be concentrated into a focus, meaning that some sort of reflective surface is necessary. Traditional mirrors, being made of solid glass, are heavy, cumbersome, and difficult to polish. Are there any other solutions?

One idea is to use a reflective liquid, mercury for example. When rotated, a liquid surface takes on the shape of a parabola under the combined action of gravity and centrifugal force [8].[†] Such a solution is inexpensive, dispenses altogether with tedious polishing, and is easy to set up. The drawback is that the mirror axis can only be vertical, limiting observations to objects near the zenith. But simplicity and low cost make the

[†] This is a classic problem in basic physics. The focal distance of the parabola is given by $F = g/2\omega^2$, where g is the acceleration of gravity and ω is the angular velocity of rotation.

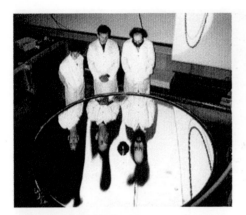

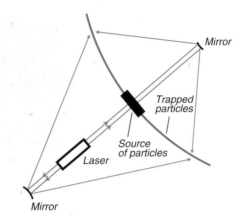

Liquid mirror 2.5 m in diameter turning at a speed of one revolution every 10 s. Credit: Borra/Laval Univ.

concept a promising one for very large diameters. As a matter of fact, a liquid mirror 100 m in diameter has been proposed for installation on the Moon.

Another approach is to forego the mirror's physical structure altogether, retaining only the reflective coating. After all, the only purpose of a massive telescope with a mirror weighing several tons and a mount weighing hundreds of tons is to support a thin layer of a reflective metal one ten-thousandth of a millimeter thick and weighing no more than a few tens of grams.

One such idea, proposed in the 1970s [36], uses light waves to trap metallic dust in a thin layer shaped like a parabola. In the original proposal, the light emitted in opposite directions by a laser is folded back on itself by two small mirrors; the reflected beams interfere to produce a series of parabolic fringes. Diffractive and scattering forces cause metallic particles to be attracted by the dark fringe, forming the reflective surface of a giant mirror. This concept is currently under study by NASA for the large mirrors in space needed to detect exoplanets.

221 Where are the best astronomical sites?

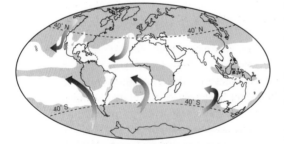

In blue–gray, the regions of the globe where over 25% of the sky is covered with clouds 50% of the time (yearly average). Arrows indicate the location of the main cold water ocean currents.

What makes a good observatory site? Many cloudless nights and, for good images, minimal atmospheric turbulence.

For cloudless nights, the favorable zones are near the tropics, between 10° and 35° north and south, as shown in white on the adjacent map.

As far as image quality is concerned, degradation is due to temperature fluctuations in the air caused by turbulence (Q. 218). The effect is strongest near the ground. During the day, solar radiation warms up the land, creating updrafts, and

Location of major world optical observatories. Many observatories in Europe, the USA, China, India, and Japan are not indicated because light pollution or mediocre image quality now prevents them from making cutting-edge observations. Antarctica is a promising site for infrared observations because it is so dry.

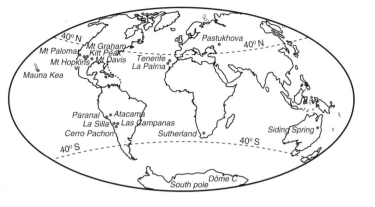

The Mauna Kea observatory, at an altitude of 4200 m on the summit of a dormant volcano on the island of Hawaii, is currently the largest international observatory. Many countries, including the USA, Canada, Japan, the UK, and France, are represented there.

at night the land cools down by radiating heat towards the sky, engendering downdrafts. These movements of air masses at different temperatures give rise to strong variations in temperature that are then exacerbated by wind action. The upper limit of these effects is the *inversion layer*,[†] which is located at an altitude of about 1000 m and is usually indicated by the presence of a cloud bank or layer of smog. Observatory sites must be located above this layer. Furthermore, the air above that layer must also be as unperturbed as possible, meaning that the high-altitude winds must be weak (no strong jet stream, in particular) and there are no large obstacles (mountain range) upwind to disturb the airflow.

In practice, such sites are found in the subtropics (i.e. between 20° and 35° latitude in both hemispheres) where upper winds are relatively weak, and on high peaks located on islands far from large land masses or on the edges of continents facing into the prevailing winds. Cool ocean temperatures (resulting from cold currents, for example), also help by lowering the inversion layer. Such exceptional sites are extremely rare, being essentially limited to the island of Hawaii, the Canary Islands, and the coast of Chile (the California coast fulfills the requirements, but now suffers from light pollution). Most of today's large observatories are found in these few regions.

[†] That is, the layer of air at which the temperature, which had been decreasing with altitude, begins to rise instead.

222 What are the advantages of observing from space?

From the very beginning of the space age (1950s), astronomers have been knocking at the doors of military and civil space agencies, trying to gain access to sounding rockets and satellites. This is because space is an ideal vantage point for astronomical observations:

1 Free from atmospheric turbulence (Q. 218), telescopes in space can produce almost perfect images, resulting in much greater resolution and sensitivity.
2 Free from atmospheric absorbtion, radiation from celestial objects can be captured over the full electromagnetic spectrum, whereas observers on the ground are limited to a few narrow "windows" (Q. 228).
3 The sensitivity of infrared observations is greatly improved because the telescope can be cooled to a very low temperature, eliminating the parasitic background created by its own heat (this cannot be done on the ground because the optics would be covered with frost).

Space has technical advantages as well as scientific ones, including the absence of gravity and the very low level of mechanical and thermal perturbations. Wind and gravity, two major factors on the ground, make it difficult to build very large optical telescopes (100 m+) there, whereas no fundamental physical limitations seem to apply to the size of space telescopes.

In that case, why not make all astronomical observations from space? The answer is simply cost. For the same collecting area, a space telescope is 10 to 100 times more expensive than one on the ground. This is true of operation costs, too. As an indication, the total cost (operation and amortization over the predicted lifetime) of an 8 m telescope on the ground is of the order of $8,000 per hour, while it is about $180,000 per hour for the 2.4 m Hubble Space Telescope (but worth every penny in terms of scientific returns!).

A further drawback of space telescopes is that human intervention for maintenance or upgrading is either impossible or very costly. The Hubble Space Telescope is currently the only one to be serviced in space, and that at a cost of more than $200 million per servicing mission.

223 What are the main space observatories?

Since observations from space are unaffected by atmospheric absorption, permitting access to the full electromagnetic spectrum, most space "observatories" and "missions"[†] since the beginning of the space age have been dedicated to covering the wavelengths that are inaccessible from the ground.

[†] In space astronomy, an *observatory* is a major facility open to the entire scientific community, while a *mission* has a more specialized observing program and is typically reserved for a single scientific team. The spacecraft employed for planetary explorations are referred to as *space probes*.

The first true space observatory was the Orbiting Astronomical Observatory, a set of three
satellites, OAO-1, -2, and -3,
launched by NASA between 1966
and 1972. OAO-1 was a failure, but OAO-2 and -3 (renamed
Copernicus) functioned for several
years, observing in the infrared
and ultraviolet with 20 to 80 cm
telescopes. Next came the IUE

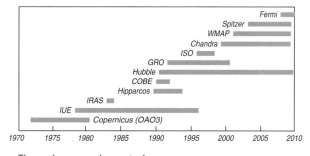

The main space observatories.

(International Ultraviolet Explorer) launched by NASA on a geosynchronous orbit in
1978 and operated jointly with the European Space Agency and the UK. Equipped with
a 45 cm ultraviolet telescope, it functioned for 18 years, one of the longest lives ever
for a satellite. Then came IRAS (InfraRed Astronomical Satellite), launched by NASA in
1983, which did an all-sky survey in the infrared with a 60 cm telescope cooled to
better than 10 K. These first successes blazed the path for numerous observatories and
missions in subsequent decades.

During the 1980s, NASA undertook a systematic exploration of the entire electromagnetic spectrum with its "Great Observatories" program. The purpose was to cover
the domains of gamma rays, x-rays, ultraviolet, visible, and infrared with major space
observatories conceived for long lifetimes and open to scientists the world over.

The first of these observatories, the Hubble Space Telescope (HST), was launched in
1990 on a low Earth orbit at an altitude of about 600 km. It is still fully operational
after almost 20 years thanks to successive visits by astronauts using the Space Shuttle,
who have made repairs and installed technical and scientific upgrades.[†] Aerodynamic
friction, even in the tenuous upper atmosphere, is not negligible for a telescope that
travels at 7 km/s. HST loses altitude, especially during periods of high solar activity
which inflate the atmosphere. It would long ago have re-entered the Earth's atmosphere
if the Space Shuttle had not periodically reboosted it to a higher orbit. HST is a 2.4 m
diameter telescope covering the visible, the infrared, and the near infrared. In spite of
its modest aperture, its angular resolution and wavelength coverage is unequaled. Its
remarkable scientific results have made it the most productive scientific satellite ever
launched.

The second observatory in the series was the Compton Gamma Ray Observatory
(CGRO), launched in 1991. In operation until 2000, it was used to carry out an all-sky
survey in gamma rays.

The third observatory, Chandra, working at x-ray wavelengths, was launched in 1999
and is still in operation.

The last of the series, Spitzer, was launched in 2003 on a heliocentric Earth-trailing
orbit (Q. 224). Equipped with a 0.85 m diameter telescope cooled to 35 K and working

---

[†] On-orbit maintenance of satellites was one of the motivations for developing the Space
Shuttle, but that proved to be much more expensive than expected, and HST is the only
satellite to be serviced regularly by it.

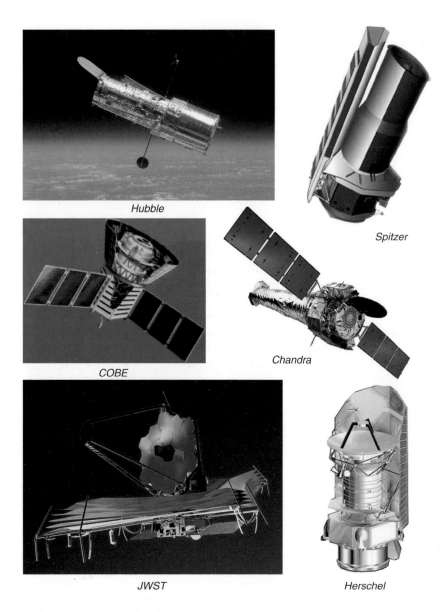

Hubble

Spitzer

COBE

Chandra

JWST

Herschel

in the infrared, it can "see" through the clouds of gas and dust that obscure galactic centers and regions of star formation. It can also detect relatively cold objects such as extrasolar planets and molecular clouds.

In parallel with these major observatories NASA launched several missions, including the famous COBE (COsmic Background Explorer) which did the first all-sky map of the cosmic background at millimetre wavelengths (Q. 133), WMAP, which redid this map at higher sensitivity, and, more recently, the Fermi Gamma-ray Space Telescope, which is studying the most energetic phenomena in the Universe such as those associated with massive black holes or exploding supernovae.

The European Space Agency (ESA), for its part, has launched several missions and observatories. These include Giotto in 1986, which made a close approach to Halley's comet, Hipparcos in 1989, which measured the position of 100 000 stars with great accuracy, ISO (Infrared Space Observatory) in 1995 which was equipped with a cryo-genically cooled 60 cm telescope, and Herschel in 2009, which carries a 3 m telescope for far infrared observations.

Several new space observatories are currently being built, the largest of which is the James Webb Space Telescope (JWST). The JWST, a joint project of NASA and the European and Canadian Space Agencies, will be launched to the second Sun–Earth Lagrange point (Q. 224) in 2013. Its main objective is to extend the study of the early Universe started by the HST. The James Webb telescope is equipped with a 6.5 m mirror passively cooled to about 50 K.

224 Which orbits are used for space telescopes?

Launching into space is enormously costly and difficult, so deciding where to locate an observatory must be a compromise between astronomical advantages and the payload capabilities of the launch systems.

Purely from the viewpoint of the energy required for launch, the "Low Earth Orbit", at an altitude of about 500 km, is the most economical. Much lower than that, drag in the residual atmosphere would cause an observatory to fall after just a few months. To benefit from the impulse supplied by the Earth's rotation, the orbit should be nearly equatorial. This is why launching facilities are preferably built at low latitudes: 28.5° for the USA (Cape Canaveral, Florida) and 5° for Europe (French Guyana). Both facilities are on the eastern side of a continent so that, in case of failure, the launcher will fall into the ocean. The Russian launch center (Baikonour) is inland at 45° latitude because that country's ocean shores are not far enough south. The period of a low Earth orbit is given by Kepler's third law (Q. 194). At an altitude of 600 km a satellite completes one orbit around Earth in 96 min.

The Low Earth Orbit is protected from cosmic rays and solar wind particles by the magnetosphere, an important advantage for astronomical observations (Q. 118). Nevertheless, there are several drawbacks:

1 Roughly half of each orbit is unusable because the Earth occults the sky field under observation.

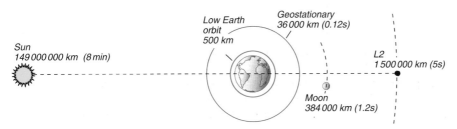

Principal orbits and locations for astronomical observatories. Distances to Earth, expressed in km and also in light-travel time, are not to scale.

2 Telescopes must be carefully protected from the strong sunlight reflected from Earth during the orbital day (even with a 4 m-long baffle in front of it, the HST cannot observe at less than 80° from the horizon of sunlit Earth).
3 Heat from the Earth nearby creates a strong parasitic background detrimental to infrared observations (an effect that can only be eliminated by being much further away from Earth).

A much better solution is to locate observatories at the second Lagrange point, L_2, of the Sun–Earth system (Q. 63), where all these problems completely disappear. Travel time from the Earth to L_2 is about 100 days, and the launch can be assisted by passing close to the Moon (Q. 78). Several observatories, planned or already in operation, are located at L_2, including WMAP, JWST, Planck, and Herschel.

Another solution is to launch an observatory at escape velocity – that is, with just enough acceleration to escape the pull of gravity (Q. 51), then let it slowly drift away from Earth. This a very economical solution from the viewpoint of launch energy since no orbital insertion maneuver is required. The observatory is then on a heliocentric orbit called an "Earth-trailing orbit." The drawback is that the observatory drifts gradually away from Earth at about 0.1 AU/year. This is because of the uncertainty about the exact amount of impulse the launching system really provides: to avoid the risk of the spacecraft falling back to Earth, the launch impulse margin must be positive. The problem is that after several years, the observatory has floated so far away that communication become impossible.

225 Would the Moon be a good site for an observatory?

Ever since the Apollo missions, the Moon has been regarded as a potentially desirable site for astronomy, combining as it does the major advantages of ground-based astronomy (stable platform, human access) with those of space (lack of atmosphere, reduced gravity). Moreover, the Moon's surface is relatively cold during the lunar night, favoring infrared observations.

The main technical problem has to do with the alternation of day and night on the Moon. Unlike free-flying observatories in high orbit which can be protected at all times from Sun and Earth radiation, sunlight reflected from the Moon's surface during the lunar day would swamp an observatory there. This, combined with the high daytime temperature of the Moon's surface (400 K), essentially precludes observations during lunar days. Half the observing time is thus lost.

But even at night, conditions are not ideal for infrared observations. Although nighttime temperatures are relatively low (100 K), they still fall short of the very low temperatures (30–50 K) obtainable passively in space. The difference is significant because thermal exchange occurs essentially by radiation and increases as the fourth power of the absolute temperature.

Finally, the logistics for building and operating an observatory on the Moon are not at all trivial. Robotic deployment of a major observatory is extremely difficult, meaning that a manned lunar base would probably have to be established before astronomical observations become possible there.

226 How is a space telescope pointed?

Action/reaction ... remember? On Earth, if you want to move a piece of furniture, no problem: you just brace your feet against the floor and push. And a motor rotates a telescope by bracing against a part of the mount that is fixed to the ground. But in space? – nothing to brace against!

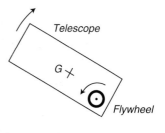

To change the orientation of a body in space, a torque must somehow be applied around the body's center of gravity. There are several ways of doing that, the most common being ejection of mass (gas jets, for example), as in a jet engine, and momentum transfer.

Gas jets are often used for fine adjustments in satellite orientation or of space vehicles such as the Space Shuttle. For astronomical observatories, however, jets have several drawbacks: they have limited lifetimes (depending on the amount of gas that can be carried), and they are a potential source of pollution for the optics. A better solution is to apply *momentum transfer*. What is that? Remember the principle of the conservation of momentum: for a body in rotation, angular momentum[†] is conserved (Q. 162). Applying this principle in space, a motorized flywheel is mounted on the body of the spacecraft. By changing its speed, the spacecraft will turn around its center of gravity in the opposite direction, keeping its total angular momentum constant. Such flywheels are called "reaction wheels." Three are needed for orienting the spacecraft in all directions, with typically a fourth one being added for redundancy.

The procedure for pointing a telescope in a new direction is then as follows. The rotation speed of the flywheel is suddenly stepped up or down. The telescope is set in motion, reaches a given speed, and continues to turn at that speed. On approaching the desired new orientation, the flywheel speed is brought back to its original value, stopping the telescope precisely on target.

From a mechanical point of view, the advantage of space telescopes is that they turn with absolutely *no friction*: no shafts, no bearings, no jittery motion ... This allows them to reach remarkable levels of pointing stability. A pointing stability of 0.1 arcsecond is trivial, and it is possible to do 10 times better. The HST, which holds the record in this domain, has a pointing stability of 7 milliarcseconds. At this accuracy, a rifle in a shooting gallery in New York City would hit the bull's eye of a target in Chicago!

227 What is an astronomical interferometer?

The angular resolution of a telescope increases with its diameter: the greater the diameter, the finer the details on its images (Q. 211). This does not mean that the collecting surface of the telescope has to be complete, however: the same resolution could be obtained with a partial surface. How so? Imagine placing in front of the

---

[†] The angular momentum of a body is the product of its moment of inertia by its angular velocity.

 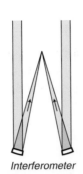

An astronomical interferometer can be thought of as a traditional telescope covered by a mask with a few holes. The image of a point source is made up of fringes within the image produced by the unmasked telescope.

Telescope   *Interferometer*

 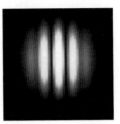

mirror of a large telescope a mask with just a few holes. Then point the telescope towards a celestial object and record the image. Now change the position of the holes and record a new image, adding it to the previous one. If the procedure is repeated until the entire surface of the mirror has been covered by the selected holes, the cumulative image will contain the same information as a single image that could have been taken with the original, unmasked telescope. The difference is that the image was obtained *sequentially* rather than all at once, and that each time an image was obtained, the collecting surface was vastly less, making for a considerable economy of mirror surface.

Individual images obtained through such a procedure result from the superposition of the light which, as it comes through each of the holes, creates interference fringes (Q. 211). Such an image is called an "interferogram."

Since nothing obliges the openings in the masked mirror to belong to the same physical optical surface, they can be in the form of separate telescopes, with their output beams combined into a common focus.

Such a system is called a "stellar interferometer" at optical wavelengths (visible, infrared), and a "radiointerferometer" at radio wavelengths. The procedure for obtaining an image by cumulating the data obtained with each configuration of individual telescopes is called "aperture synthesis."

The advantage of interferometers is their very high angular resolution. Widely separated individual telescopes provide a vastly better angular resolution than that of a single telescope with the same amount of collecting surface. It is the resolution of an immense single telescope with a diameter equal to the maximum separation of the interferometer's individual telescopes. The VLTI interferometer of the European observatory in Chile (ESO), for example, has the resolution of a 200 m telescope. In the radio domain, the American VLA (Very Large Array) in New Mexico extends over 20 km, which gives it a resolution of 0.05 arcsecond, 10 times better than traditional optical telescopes can provide.

There is a price to pay, of course. Since the collecting area for each image is much smaller than that of a corresponding full-aperture telescope, such images are very "noisy." Interferometers have high resolution but poor sensitivity. They can only observe bright objects.

228 How does a radio telescope work?

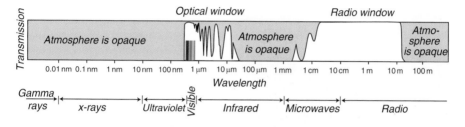

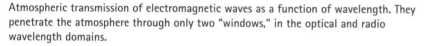

Atmospheric transmission of electromagnetic waves as a function of wavelength. They penetrate the atmosphere through only two "windows," in the optical and radio wavelength domains.

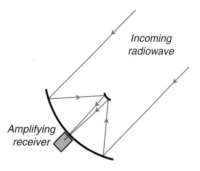

Since the late nineteenth century there was a suspicion that celestial objects emitted radio waves, but the actual emissions were not discovered until 1932 – and then just by chance, not by an astronomer but by an American radio engineer, Karl Jansky. Furthermore, this cosmic broadcast was at first only of interest to radio engineers, Grote Reber in particular (Q. 244). It was only after the development of radio electricity during World War II, the invention of radar in particular, that radio waves captured the attention of astronomers. This new science immediately blossomed, and the years 1950 to 1970 saw the construction of a profusion of large radio telescopes, among them a 76 m at Jodrell Bank, UK (1957), a 90 m at Greenbank, West Virginia (1962 – it collapsed in 1988), a 40 × 200 m at Nançay, France (1963), a 300 m at Arecibo, Puerto Rico (1963) and a 100 m in Effelsberg, Germany (1972).

The radio domain is important because it is one of our two "windows" onto the Universe in our otherwise opaque atmosphere. It extends from a centimeter to a few hundreds of meters in wavelength.

Although radio waves are not fundamentally different from light, the energy they transport is much less. To perceive the very faint radio emission of astronomical sources requires large collecting surfaces and amplification of the received signals.

Radio telescopes can be of various shapes, primarily depending on the wavelengths being studied. Some are simple arrays of antennas resembling TV antennas, but most work on the principle of optical telescopes: a large parabolic reflector concentrating

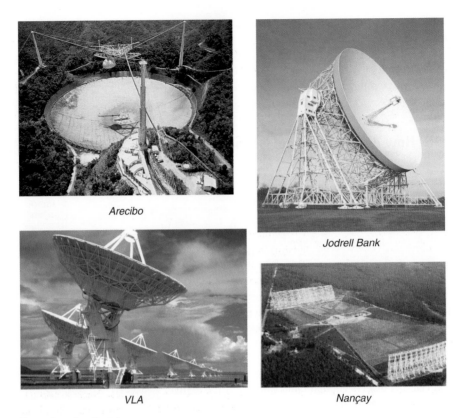

Arecibo

Jodrell Bank

VLA

Nançay

Examples of radio telescopes.

the incident radio waves at a focus, either directly (prime focus) or after reflection on a secondary mirror (Cassegrain focus). The signal is captured by a small antenna at the focus and sent to a receiver where it is amplified.

A radio telescope dish must be much larger than the mirror of an optical telescope, not only because the radiation received is very weak, but also to be able to pinpoint where it is coming from. This is because the resolution of any telescope, optical or radio, depends on the ratio of the observed wavelength to the diameter of the telescope (Q. 212). Since radio waves have very long wavelengths, the diameter of a radio telescope must be very large to avoid seeing the sky all blurry. At a wavelength of 21 cm, the 300 m Arecibo telescope in Puerto Rico, the largest in the world, cannot distinguish celestial sources smaller than the Moon. And at that same wavelength, a radio telescope would have to have a diameter of 40 km to distinguish details that can easily be seen by an amateur optical telescope.

The disappointing resolution of radio telescopes quickly prompted the construction of radio interferometers (Q. 227) spread over several kilometers, and even over half the Earth, like the American VLBA (*Very Long Baseline Array*) or the European EVN (*European VLBI Network*). These very long baseline interferometers have a resolution on the order of one thousandth of an arcsecond, which is a hundred times better than that

Location of the individual telescopes of the very long baseline radio interferometers: EVN in red and VLBA in green.

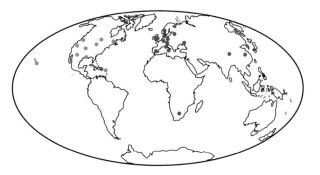

of the best optical telescopes. Unlike optical interferometers, the individual telescopes of radio interferometers do not need to be interconnected. They simply need to be aimed at the same time at the same field, and the recorded signals are combined post facto.

A radio telescope does not produce images directly. The sky field of interest is scanned by making a series of successive telescope pointings, and the image is reconstructed using the signal recorded at each of the pointings.

Because of the longer wavelength of radio waves, radio telescope mirrors do not need to be as accurate as those in the optical domain. They do not even need to form a continuous surface. Most are actually composed of wire mesh or of perforated plates, so as to reduce weight and wind pressure. The same principle is used in a microwave oven: its glass door is covered by a metal screen with small holes. These are much smaller than the 12 cm wavelength of the microwave radiation used to cook food, so the screen *reflects* the microwaves back to the interior of the oven and nothing leaks through.

Radio emission from celestial sources is extremely weak compared to that of TV broadcasts and cell phones. Radio telescopes are thus preferentially placed far from urban regions, and in valleys that shield them from "radio pollution" – not on mountaintops like their optical cousins. The American Green Bank radio telescope in West Virginia is even located in a special $35\,000\,km^2$ "radio quiet zone," set aside by the Federal Communications Commission (FCC).

229 What can we learn from observations at radio wavelengths?

Observations at radio wavelengths are highly informative because they allow us to see things that are not detectable in the optical. Radio waves pass easily through gas and dust clouds, so stars or galaxies that would be invisible to an optical telescope can be seen at radio wavelengths.

Certain regions of the Universe contain molecules that absorb or emit only at radio wavelengths. This is the case with neutral hydrogen, which emits at a wavelength of 21 cm. By observing this emission line, velocity and temperature maps can be made of these regions.

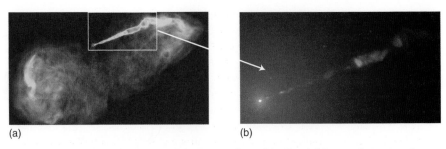

(a)   (b)

Example of how observations at radio wavelengths and in the visible complement each other: (a) the Galaxy M87 with its jet of subatomic particles seen in radio waves by the VLA, and, (b) the same jet seen in the visible by the HST. The radio image reveals the presence of a giant structure, undetectable in the visible, whose emissions in radio and x-rays are produced by jets of particles emanating from the Galaxy. Credit: NRAO/NSF/STScI.

In addition, some objects emit strongly in the radio domain but weakly in the visible. This is the case of pulsars (neutron stars spinning at fantastic speeds), of quasars (young galaxies that are so distant that they appear as a single star, but which are among the most energetic objects in the Universe), and of radio galaxies. All these objects can only really be detected in the radio domain.

For all these reasons, radioastronomy is an essential complement to optical astronomy in furthering our understanding of the Universe.

230 What is a submillimeter telescope?

Artist's view of the ALMA submillimeter interferometer in compact configuration. The individual antennas, which are 7–12 m in diameter and weigh approximately 100 tons, are repositioned by a custom–designed transporter. Credit: ESO.

The submillimeter domain, occupying the band between infrared and radio waves, has the particularity of not being absorbed by interstellar dust. It is the ideal wavelength to study the regions of star formation and galactic nuclei.

Submillimeter tele- scopes are similar to radio telescopes, except that their mirror surfaces must be much more precise. An additional difficulty is that submillimeter waves are more absorbed by water vapor than are radio waves. Submillimeter telescopes must be located in extremely dry sites such as Antarctica, high mountains, or stratospheric balloons. Two submillimeter telescopes, 10 and 15 m in diameter, have been installed at the Mauna Kea observatory in Hawaii at an altitude of 4200 m, and ALMA (Atacama Large Millimeter/submillimeter Array), a gigantic array of 66 antennas spread over a radius of 10 km, is currently being built on a high Chilean plateau at an altitude of 5000 m. ALMA, which is a joint project of the USA, Canada, Europe, Japan, and Taiwan will be completed in 2012.

231 What does an x-ray telescope look like?

An x-ray telescope cannot use a traditional mirror. X-rays have such high energy that instead of bouncing off its surface, they would penetrate it and stop there – much like a gun bullet slamming into a wall. However, a way to deflect and focus x-rays was found in the 1950s by the German physicist Hans Wolter, while trying to build an x-ray microscope: x-rays can be reflected if they hit a metal surface at a very shallow angle – grazing it – just as a stone skips over water. Since the deflection angle is small, the focal length of such a mirror is very long. To reduce the length of the telescope and also improve image quality, two successive mirrors are used, a paraboloid followed by a hyperboloid with the same focus, as in the Ritchey–Chrétien combination (Q. 206). And to increase the collecting area, several of these mirrors are nested like Russian dolls.

Imaging with an x-ray telescope is different from medical radiography. In medical applications, the part of the human body to be studied is placed between an x-ray source and a photographic film. Bones, which are denser than muscle and skin, absorb more of the x-rays and appear lighter on the film. An x-ray telescope, on the contrary, forms an image of a celestial object, just like a regular telescope.

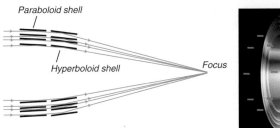

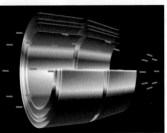

Grazing incidence mirrors for x-rays.

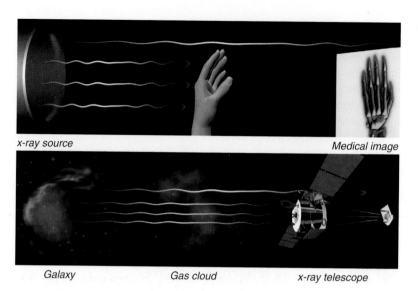

x-ray source   Medical image

Galaxy   Gas cloud   x-ray telescope

Unlike medical radiography, an x-ray telescope forms an image of the x-ray source, not that of the intervening objects. Credit: NASA.

232 What can be learned by observing at x-ray wavelengths?

The x-ray Universe is very different from the visible one as x-rays are produced only by matter that has been heated to temperatures in the millions of kelvins. And this occurs only in the presence of extremely powerful magnetic or gravitational fields, or during explosions. See Q. 156 for a view of the Galaxy in x-rays.

(a) image of a galaxy cluster taken in visible light and, (b) in x-rays (false color). The x-ray image reveals that the cluster is enveloped by a cloud of gas at a temperature of 40 million K. Credit: McNamara/NASA.

(a)   (b)

233 How does a gamma ray telescope work?

Gamma rays are high-energy photons. Because they interact strongly with the atmosphere and are absorbed by it, gamma ray astronomy must be done from space. But the rays are so energetic that they cannot be captured by mirrors; rather than being reflected, they pass right through them. They are detected by the flashes they provoke when striking certain crystals or liquids. The light or electrons this produces can be visualized with a light detector or by means of a spark chamber such as those used in particle physics. This permits their direction and energy level (i.e. wavelength) to be determined.

The first important gamma ray observatory was the *Compton Gamma Ray Observatory* (CGRO), which was used to study the origin of the mysterious "gamma ray bursts." These powerful flashes of gamma radiation were discovered during the Cold War by American military satellites monitoring Soviet nuclear tests. The energy they emit is fantastic, as much produced in 10 s as the Sun has produced in all the time of its existence. They are thought to be cataclysms associated with supernovae or merging black holes in distant galaxies.

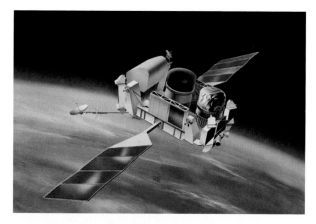

CGRO, launched in 1991. The two cylinders are the scintillation detectors. Credit: NASA.

In 2002, the European Space Agency launched *Integral* (International Gamma-Ray Astrophysics Laboratory), a satellite designed to make fine spectroscopic analysis of gamma ray sources. In 2008, NASA launched *Fermi*, the successor of CGRO.

234 How are gravitational waves detected?

When a massive body moves or changes shape, it alters the fabric of space-time (Q. 144), creating a "gravitational wave" – just as a moving boat or a pebble tossed into a lake creates a wave in the water. Gravitational waves, traveling at the speed of light, affect physical bodies in their paths – just as water waves make a piece of floating wood bob up and down on encountering it.

Although predicted to exist by the theory of relativity, no one has ever directly observed a gravitational wave. The effect is a very weak one and the only hope of detecting one would be in association with enormous moving masses such as massive binary stars, compact objects rotating around black holes, exploding supernovae, or black holes in formation. Even in such cases the effect would necessarily appear tiny to

Artist's view of LISA. Credit: NASA/ESA.

us, because the amplitude of gravitational waves decreases with the distance traveled. It is estimated that, if one should reach Earth, its passage would displace an object by less than the width of a single atom.

If one were detected, however, it would be a capital event. Not only would this be another confirmation of the theory of relativity, it would also open a new "window" on the Universe, as happened with radio waves and x-rays. Several gravitational wave observatories are now either in preparation or already in operation, all based on the idea of observing the minute displacement of a proof mass during the passage of a wave. Notable among these, on the ground, is LIGO (*Laser Interferometer Gravitational-Wave Observatory*), which has synchronized detectors 3000 km apart, located in the states of Washington and Louisiana. In space, NASA and the European Space Agency are collaborating on a triple satellite system, LISA (*Laser Interferometer Space Antenna*), scheduled for launch in 2011. LISA's satellites, each containing a proof mass floating freely inside, will be placed at the corners of a triangle 5 million km/side, and the distances between the proof masses will be measured by laser interferometry.

235 How are neutrinos detected?

Neutrinos are elementary particles that carry no electric charge. They have no – or very little – mass and travel almost at the speed of light. Neutrinos are created during nuclear fusion reactions in stars and in cataclysmic events such as supernova explosions. Billions of neutrinos originating in the Sun pass through our bodies every second.

With no electric charge and almost no mass, they rarely interact with matter and so are extremely hard to detect. Enormous tanks of water are used for that purpose: when by chance a neutrino smashes into an atom, it leaves behind either a radioactive body or a luminous ray. The tanks are buried deep underground (in disused mines) so as to be sheltered from cosmic rays which produce the same effects.

Neutrino astronomy is just in its infancy, but it is opening a new "window" onto the cosmos. Everything that we know about the furthest reaches of the Universe has been learned thus far thanks to *electromagnetic radiation*. Neutrinos are not a form of radiation; they are a form of *matter*.[†] And that matter brings us information about the Universe almost as speedily as electromagnetic radiation.

236 How is observing time allocated in a modern observatory?

Observatory schedules are programmed months in advance. Proposals for observations are typically examined by a committee once a semester, and are selected strictly on the basis of scientific merit. When a telescope is in great demand (HST, for example), only about one proposal in four or five can be accepted. Research in astronomy is a science without borders: observations are usually open to all astronomers, regardless of nationality – what is called an "open skies" policy.

[†] Cosmic rays are matter, too, but since they are composed of electrically charged particles, their paths deviate in the magnetic fields of Earth and of the Galaxy, making it impossible to discover their point of origin.

For ground telescopes, an observing run can last anywhere between a few hours and a few nights (or days, for radioastronomy), and for a space observatory, from one hour to hundreds of hours.

Ground telescopes are now becoming so automated that astronomers sometimes do not even need to leave their home offices to make their observations; unless very special measurements are required, the observatory astronomers and technicians can execute their program and transmit their data to them via the Internet.

Amateur astronomy

237 Interested in amateur astronomy? What are the first steps?

Astronomy is not just for professionals – the sky belongs to everyone! Amateur astronomy is a fascinating hobby, running from the simple pleasure of gazing at the night sky, to learning to appreciate the phenomena and mysteries of the Universe, all the way to making quasi professional observations. Perusing books and magazines on astronomy, even studded with spectacular photographs, can never match the emotional impact of engaging directly with the heavens.

Pastime or true passion, here is an activity that is within almost everyone's means. Even the most sophisticated amateur astronomers spend significantly less money on their hobby than do many sportsmen, boating enthusiasts, and golfers. It is important to start out on the right foot, though. If you just rush in and buy a cheap "toy" telescope, you will quickly be disappointed and lose interest. On the other hand, if you acquire the biggest, most expensive instrument on the market, you are likely to find yourself overwhelmed.

The best way to start is to observe the sky with the naked eye from a dark site using a star chart or one of the new handheld devices using GPS technology (such as SkyScout). Once you have learned to find your way in the sky and are familiar with the major constellations and planets, you can move up a step, to binoculars. Standard models, 8 × 40 or 7 × 50,[†] are good choices. They are relatively inexpensive, convenient to use, and offer a generous field of view, making it easy to locate objects. Then you can savor the beauty of the moons of Jupiter, the Andromeda Galaxy, the Pleiades, the Orion Nebula, several star clusters ... and our spectacular Moon.

If your interest grows, a small telescope would be your next step. Amateurs have to accept the fact that, even with a good quality amateur telescope, visual observations can never hope to equal the spectacular professional images produced with large telescopes and long exposures. Nonetheless, the excitement of being in direct contact with the Universe, that strange feeling of seeing celestial objects suspended in space, is one of the incomparable pleasures that visual observations with a telescope can provide. If you really get hooked and want to try astrophotography, there are now some impressive digital imagers on the market that will allow you to produce magnificent astro-images,

---

[†] Binoculars are described by two numbers, the first indicates the magnification, the second the diameter of its objective lenses in millimeters. The larger the latter number, the more light the binoculars gather, but also the more cumbersome and the heavier. Wide field 8 × 40 and 7 × 50 models are an excellent compromise. Even if you plan to acquire a telescope later on, binoculars are always handy to have, as you can use them to identify objects before observing them through the telescope. Mounting binoculars on a sturdy tripod stabilizes viewing conditions.

somewhere between science and art, but this calls for no small measure of dedication and technical mastery.

Guidance on choosing a telescope can be found in Qs. 238 and 239. Before buying, take the time to do a bit of research: read some relevant books and magazines, check out the telescope manufacturers' websites, and visit specialty shops. If possible, attend night-sky observing sessions or "star parties" organized by astronomy clubs (Q. 250) where you can compare various instruments for size and ease of use and get the benefit of advice from other amateurs.

238 Which telescope should you choose?

There is no one "best" telescope to suit everyone. As you go about deciding what is best for you, you must keep in mind four main criteria: your observation priorities, your budget, and your requirements concerning the instrument's ease of use and its transportability.

You will have to begin by choosing between a refractor (with lenses) and a reflector (based on mirrors) (Q. 202). If you are mostly interested in bright objects like the Moon, the major planets, double stars, star clusters, and the brightest Messier objects (Q. 242), a refractor is your best option. With no multiple reflections to diffuse light and no secondary mirror to obstruct the beam, the refractor offers better contrast and produces clear images without spikes. Moreover, its throughput (90%) is generous compared with that of most reflectors (75%); an 8 cm (3 in) refractor may produce images as luminous as a 15 cm (6 in) diameter reflector. It must be of good quality, however, not the kind you find in discount stores, and that means an instrument that will be more costly than an equivalent reflector. The performance of a refractor is characterized by the diameter of its objective lens and the quality of its *chromatic correction*, i.e. compensation for the dispersion of light in the glass (Q. 210). The larger the diameter of the refractor, the more luminous the image and the fainter the object you will be able to see. The magnification is not a performance factor – the eyepiece determines that. Most common refractors have diameters between 60 and 100 mm (2.5–4 in).

If you are keen to observe fainter objects like nebulae and galaxies, you might prefer a reflector. For this type of object, it is important to collect as much light as possible, which means having a large diameter telescope. And for the same diameter, reflectors are much less expensive than refractors.

A wide range of amateur reflectors is available, from the simplest Newtonian to the many Schmidt–Cassegrain designs.

Newtonian reflectors, based on the configuration invented by Isaac Newton (Q. 204), are constructed around a concave mirror, the primary mirror, and a small flat mirror that redirects light out of the top of the tube. Its simple optical

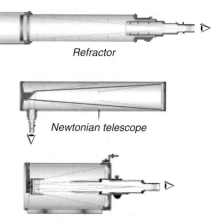

Refractor

Newtonian telescope

Schmidt–Cassegrain telescope

design makes it inexpensive. However, beware of the "cheapies" you find in discount stores. They are destined to be huge disappointments as astronomical instruments because of their poor optics and fragile mechanisms. The main disadvantage of the Newtonian is that having only one curved mirror, off-axis aberration cannot be corrected (Q. 206). To reduce this effect, the telescope needs a long focal length, making it relatively bulky. Even with a fairly long focal length, the usable field of view remains small and does not allow the observation of extended objects or astrophotography. Other disadvantages are that the optics fall out of adjustment easily and the mirror, exposed to the ambient air, is subject to dust buildup.

The Schmidt–Cassegrain (Q. 241) and its Maksutov–Cassegrain variation correct all these defects: they are compact, offer a good-sized field of view, stay aligned and collimated, and their mirrors are generally protected. Thus, they are much more user-friendly. Unfortunately, they are also more expensive. Nevertheless they come close to being ideal instruments for dedicated amateurs, in particular those interested in astrophotography.

A reflecting telescope is typically characterized by two numbers: the *diameter* of the primary mirror and its *effective focal length*. The effective focal length of a telescope is its apparent focal length as seen from the final focus (where the eyepiece or camera is placed). For example, a 115/900 (4.5/35) telescope has a diameter of 115 mm and an effective focal length of 900 mm. The most common amateur reflectors have diameters between 10 and 20 cm (4 to 8 in). The cost of a reflector depends mainly on its diameter and not its focal length. For a given design (Newtonian, Schmidt–Cassegrain, or Maksutov–Cassegrain), the cost increases with the square of the diameter.

Whether you choose a refractor or a reflector, it is important to have a good set of eyepieces, with focal lengths ranging from about 5 to 40 mm to deliver different magnifications. You just need a few, but it is good to include one that provides a wide field of view, 80° or more. Be forewarned that the cost of a small set of high-quality eyepieces may equal that of the telescope itself.

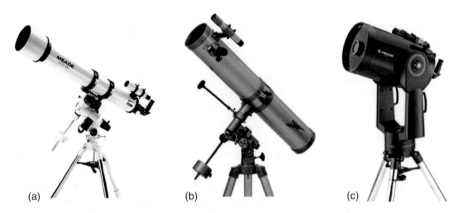

(a) a refractor, (b) a Newtonian reflector (both with equatorial mounts), and (c) a Schmidt–Cassegrain telescope on a motorized alt–az mount. Credit: Meade and Celestron.

Your second important choice is of a mount, the structure and mechanism that allow the telescope to be oriented and to track celestial objects. There are two types of mount: alt-azimuth and equatorial (Q. 203). The alt–az mount is simpler but requires simultaneously moving the telescope along two axes to track objects as the Earth rotates, and doing this manually is a bit awkward. On more sophisticated instruments, these two movements are under computer control. However, this type of mount is not suitable for long photographic exposures because the field rotates.

In the equatorial mount, the telescope needs to be moved around only one axis (the polar axis), and this at a constant speed (a type of movement that is easily motorized). In addition, the field does not rotate, which permits long exposures.

There is no major price difference between the two designs. The equatorial mount has the disadvantage of being heavier when a counterweight is used to balance the weight of the telescope. Whatever type of mount you choose, its construction should be sturdy and of good mechanical quality to ensure accurate pointing and stability during observations. For a relatively reasonable price, motorized mounts can be equipped with systems called *Go-To* that automatically point the telescope at a given object if you simply type its name on a computer keyboard. This is wonderful for beginners or those with limited time to devote to observing.

Remember, finally, that to observe faint objects you need a dark sky, meaning that you must set up your instrument at least 40 km from any large city, and that may mean packing it up and driving it around, hence the importance of transportability. The bulkier and heavier the instrument, the more challenged you will be with its transportation and setup. You will have to find the right compromise between performance and portability.

239 What can be seen with an amateur telescope?

It all depends on the telescope aperture and the observing conditions – darkness of the sky and atmospheric turbulence. Here are what you can hope to view when conditions are good.

With a 60 mm (about 2 in) refractor: the surface of the Moon, the phases of Venus, sunspots (with a filter in front of the instrument – whether using binoculars or a telescope, *never* look at the Sun without a filter!), double stars separated by more than 2 arcseconds, stars up to magnitude 11.5, large globular clusters, the brightest nebulae, and most Messier objects (but with few details).

With an 80 mm (about 3 in) refractor or a 10 cm (4 in) reflector: the phases of Mercury, smaller craters on the Moon, the polar caps on Mars, the bands on Jupiter's disk and four or five of its satellites, the divisions in Saturn's rings, double stars separated by more than 1.5 arcseconds, stars up to magnitude 12, dozens of globular clusters, several planetary nebulae, galaxies, and all Messier objects (Q. 242).

With a 15 cm (6 in) reflector: all the previous objects in greater detail, the centers of globular clusters, and (under a very dark sky) many details in nebulae and galaxies.

With a 20 cm (8 in) reflector: details on the Moon of the order of 1 km, several asteroids, stars up to magnitude 14, the core of globular clusters, and (with a very dark sky) exquisite details in nebulae and galaxies.

Visual observation has its limits. Although the eye is very sensitive (equivalent to an ISO 800 film) once dark-adapted, it can integrate (accumulate and process)

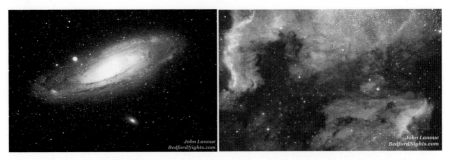

A 15–20 cm (6–8 in) telescope can produce very fine astronomical images, such as these of the Andromeda Galaxy and of a nebula, with an exposure time of 1–2h, captured by an experienced amateur astronomer. Credit: J. Lanoue, BedfordNights.com.

low-intensity light for a few seconds at most before sending an image to the brain [13]. After that it resets itself. If, instead, a photographic camera and a motorized mount are used, exposures only a few minutes long will reveal details invisible to the naked eye. A simple webcam modified for long exposures can also give good results.

The electronic age has revolutionized amateur astrophotography. Not only are digital detectors much more sensitive than photographic emulsions, they allow multiple exposures. With the proper software, poor-quality individual images can be removed and the best ones added together. The recent appearance on the market of low-cost electronic detectors (like charge couple devices – CCDs) specialized for astronomy has provided amateurs with powerful tools to produce spectacular color images, rivaling in beauty those made by professionals.

240 What is a Dobsonian telescope?

A Dobsonian telescope is more a design approach than a new telescope type. Its goal is to provide a large diameter telescope, easy to use and transport, at the lowest possible cost. The idea originated with the American amateur astronomer John Dobson in the 1950s. As a monk in a Hindu temple in California, Dobson was given the task of reconciling astronomy and the teachings of Hinduism. This led him to construct his own telescopes, and he became interested in bringing the wonders of the cosmos to as large a public as possible.[†]

The Dobsonian telescope is actually a Newtonian with a fairly large primary mirror (20–30+ cm) on a manually operated alt–az mount. In John Dobson's original version, the mirror was ground out of a ship's porthole and sat at the bottom of the tube on a piece of carpet. The tube was made from a length of Sonotube® (a cardboard tube normally used as formwork for concrete columns), and the mount was built of plywood, with the two rotation axes set on Teflon pads. The whole thing could be put together

---

† After leaving the monastic life, Dobson would often set up his telescope on the sidewalks of San Francisco and invite passers-by to take a look at celestial objects. He is the founder of the Society of Sidewalk Amateurs (www.sidewalkastronomers.us). Its many volunteer members around the world perpetuate the tradition.

by an amateur with no special skills and at minimal cost. Since then, innovative and creative amateurs have improved on the design, but the principle remains the same: low cost and easy assembly. Commercial versions are now available, including some with electronic drives and *Go-To* capabilities.

In addition to its low cost, the Dobsonian, with its large diameter, gives access to faint objects of the deep sky that more expensive traditional telescopes cannot attain. Its disadvantage is that, in its basic, unmotorized version, it has to be continuously adjusted in order to track objects. And as for other amateur alt-azimuth telescopes, long photographic exposures are impossible because the field of view rotates (Q. 203).

241 What is a Schmidt–Cassegrain?

A Schmidt–Cassegrain telescope is composed of a spherical primary mirror, a thin lens (called a plate) that corrects the spherical aberration, and a secondary mirror mounted on the plate and focusing the light beam behind the primary mirror, as in the Cassegrain design (Q. 206). The primary-mirror/plate is not a pure Schmidt configuration (Q. 208) because the plate is not at the mirror's center of curvature, but near its focus. This is acceptable because the telescope is not meant to produce a large field of view.[†]

The Schmidt–Cassegrain is popular with amateurs because it is even more compact than a regular Cassegrain (because a faster primary, typically $f/2$,[‡] can be used), and its tube is closed, thus protecting the primary mirror from dust. Another advantage is that images are free of diffraction spikes, since the secondary mirror is mounted on the plate instead of on supporting beams (Q. 9). Its drawback is higher cost because the complex shape of the plate is difficult to manufacture.

The Maksutov–Cassegrain is a design variation in which the correcting plate is replaced by a strongly curved meniscus. This design is less expensive than the Schmidt–Cassegrain because its optics are made solely of spherical surfaces and thus are easy to machine produce. Its disadvantage is its long effective focal length ($f/15$ instead of $f/10$ for a Schmidt–Cassegrain), which makes it less suitable for astrophotography.

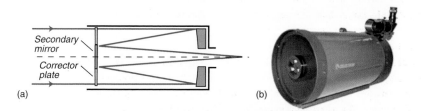

(a) Optical diagram of the Schmidt–Cassegrain telescope shown at (b), where one can see the correcting lens carrying the secondary mirror. Credit: Celestron.

[†] This optical combination was invented by Roger Hayward (1899–1979) during World War II for nighttime infrared photography.

[‡] For a definition of the focal ratio, see Q. 204.

In a variation of the Maksutov–Cassegrain, the large thin meniscus at the telescope entrance is replaced by a small thick meniscus immediately in front of the secondary mirror. The tube is open, reducing problems with condensation, but protection from dust is lost and images suffer from diffraction spikes.

242 What are the Messier objects?

Charles Messier.

The Messier objects are the delight of amateur astronomers.

The French astronomer Charles Messier (1730–1817) was a dedicated comet hunter. Systematically searching the sky for non-stellar objects, he found a number of diffuse objects that he soon realized could not be comets because they did not move. To help other comet hunters in their searches, he published a list of all such diffuse objects that he and his assistant Pierre Méchain had found. The last edition of his *Catalog of Nebulae and Star Clusters*, published in 1784, lists the brightest "nebulae" and globular star clusters in the sky. At the time, the term

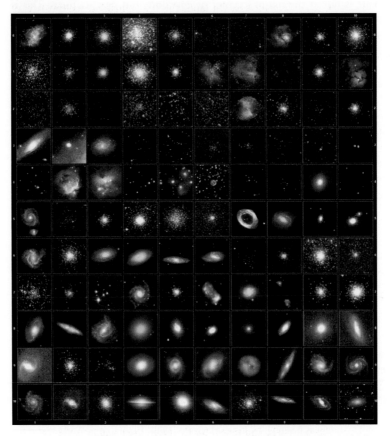

The 110 Messier objects: the most spectacular star clusters, galaxies, and planetary nebulae of the boreal sky that can be seen with a small telescope.

nebula was applied to any object that appeared to be a cloud of gas, but we now know that many of these objects are actually galaxies other than our own (Q. 157). Seen through binoculars or a small telescope, these deep-sky objects are among the most spectacular that an amateur astronomer can observe. Messier made his observations from Paris, hence his catalog does not cover certain beautiful objects in the austral sky, such as the Magellanic Clouds. Still, his list can provide most observers with hours and hours of enjoyment.

243 Where are skies the darkest?

Modern life is ruining our view of the night sky. The light that floods our cities, and now more and more the countryside, produces a sky glow that washes out most of the stars. Far from major population centers, thousands of stars are visible to the naked eye. In and around them, that number drops to just about one hundred. Many city-dwellers have never seen the Milky Way, the Galaxy we live in!

To enjoy a reasonably dark sky, one has to be at least 10 km from a town of 2000 inhabitants, and 100 km from a city of 1 million. In large cities, stars can only be seen from inside a park ... or a cemetery. On land, the last remaining regions free of light pollution are high mountain areas, deserts, less developed countries, and isolated islands.

The high seas are still essentially untouched. The next time you are on a night sail, look up. You will see wonders that are hidden from the view of 95% of the citizens of developed countries.

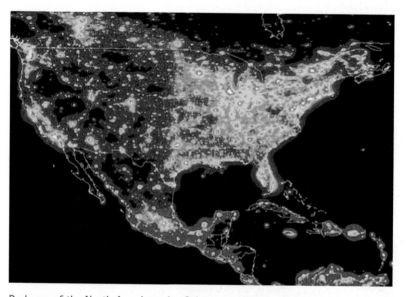

Darkness of the North American sky. Colors represent the scaled brightness of the sky with respect to maximum (natural) darkness: 10% brighter for the dark gray regions, 30% for the blue ones, 100% (that is, two times) for the green ones, and 27 times for the red regions. Credit: US Air Force, RAS and Blackwell Science.

244 What important discoveries have amateurs made?

The word amateur, French for "one who loves," did not originally imply dilettantism as it often does today. Many amateur astronomers of the past were quasi professionals.

One of the best examples is William Herschel who was a professional musician. Born in Germany, he moved to England where he pursued a musical career as an organist and composer of symphonies and concertos. After reading a book on optics written by a musical colleague, he built his own telescope and began observing the sky with his sister Caroline (also a musician). Astronomy became a passion for both. Going beyond observation of the Moon and planets that were the prime attractions for his contemporaries, Herschel directed his viewing efforts at stars and nebulae (Qs. 153 and 215). He drew the first chart of the Galaxy and showed that some nebulae were made of stars. His extensive exploration of the sky led him to the discovery of the planet Uranus. Impressive achievements for an amateur! His discovery of the new planet brought him instant fame and, at 43 years of age, he began a new career as professional astronomer, under the patronage of the king.

William Herschel
(1738–1822).

More recently, at the end of the nineteenth century, the English engineer and amateur astronomer Isaac Robert was one of the first to use the newly invented technique of photography in astronomy. His long-exposure photographs revealed details never before seen, particularly in nebulae, stunning astronomers of his day.

The pioneer of radioastronomy, Grote Reber, was also an amateur. His story starts in 1932 when it was learned that the Milky Way was a source of radio waves. A serendipitous discovery by an American engineer, Karl Jansky, who was trying to understand the origin of the static that affected long-range radio communication, it made newspaper headlines. Oddly enough, the scientific community remained completely indifferent. But radioamateur Grote Reber was intrigued. Jansky had worked with a relatively simple

Isaac Roberts (1829–1904) and the first photograph of the Andromeda Galaxy, obtained in 1888 in a 150 min exposure.

Grote Reber (1911–2002), the first radio astronomer and his 32 ft (10 m) radio telescope built in his garden.

radio antenna. Reber, in 1937, constructed the first real radio telescope using his own money and his own hands. Thereafter for many years, day and night, he spent long hours "listening" to the sky. He produced a radio map of the Milky Way and identified several discrete sources, including Cygnus A that we now know to be a system of merging galaxies. The first radio astronomer was thus an amateur, and his observations triggered the explosive growth of radio astronomy after World War II.

The dedicated amateurs who have made significant contributions to the discovery and study of planets, comets, asteroids, variable stars, and supernovae are too numerous to mention. Let us recall just a few.

There was Clyde Tombaugh (1906–97), a young American amateur who constructed a 20 cm telescope, grinding his own mirror and cannibalizing parts of old farm machinery and of his father's 1910 Buick. When he sent his impressive detailed drawings of Mars and Jupiter to Flagstaff Observatory in Arizona, he was immediately hired as an observatory assistant. A year later (1930), using the observatory's 33 cm astrograph, he discovered Pluto. Then there is Robert Evans, an Australian pastor who used his homemade 25 cm telescope to discover a record number (40) of supernovae. And what about the New Zealander Albert Jones who has beaten all records of variable star observations (more than 500 000), discovered a comet that now bears his name, and is also a co-discoverer of the supernova 1987A in the Large Magellanic Cloud? Or the Canadian-born amateur David Levy who discovered 22 comets including the famous Shoemaker–Levy that crashed on Jupiter in 1994? And Christopher Go, a young amateur from the Philippines who discovered "Red Spot Jr." on Jupiter in 2006 using a 20 cm telescope? He is now a member of a team of professional astronomers using the HST to monitor Jupiter.

245 How can planets be spotted in the night sky?

If you are out in the field and are unable to access one of the many sky maps available in magazines or on the Internet, the planets can be difficult to pick out from among the many bright stars. The only ones visible to the naked eye are Venus, Mars, Jupiter, and Saturn (Mercury requires exceptional conditions). Here is how to find them.

Since the orbit of Venus is inside that of Earth, this planet never strays far from the Sun (47° at most, i.e. about 3 h of time). Look for it in the east before sunrise or in the west after sunset. At maximum brightness it is a spectacular object – the brightest in the night sky after the Moon.

Mars, Jupiter, and Saturn, whose orbits are outside the Earth's, can appear at any time during the night, but only for part of the year. Look for them close to the track followed by the Sun during its day-time apparent trajectory (i.e. the ecliptic). Mars can be identified by its characteristic orange color. Jupiter is as bright as some of the brightest stars, while Saturn is the faintest of the visible planets.

When in doubt, there is a simple way to confirm that one is really looking at a planet: stars twinkle but planets generally do not (Q. 125).

246 Where and how can meteorites be found?

At least 50 000 meteorites weighing over 10 g strike the surface of the Earth every year. Where and how to find them? The best hunting is in relatively flat terrain, where they stand out from their surroundings. Any area with minimal vegetation can be productive, but avid meteorite hunters concentrate on deserts, dry lake beds, and ice-covered regions (professionals go to Antarctica).

Meteorites have a shiny surface and are blackish, brownish, or rusty in color. The surface of a meteorite becomes hot enough to melt during its passage through the atmosphere, and then, upon cooling and solidifying, a thin black layer called the "fusion crust" forms. Under weathering, the crust turns brownish or rust colored.

The second characteristic of a meteorite is its high iron content. Whether "metallic" or "stony," all meteorites have high iron contents and are attracted by a magnet. A meteorite hunter will often use a magnet suspended on a string to avoid bending over to pick up "imposters" that resemble the real thing. Sophisticated hunters use metal detectors.

247 Can amateur astronomers participate in serious research programs?

Astronomy has become essentially a "business" for professionals, but that does not mean that amateurs have no role to play. As a matter of fact, astronomy is one of the few sciences where an amateur can still make a significant contribution.

Times are long gone when, to be a cutting-edge geneticist, it was enough to have one's own pea garden, like Mendel, or where one could advance the frontiers of physics and chemistry, like Coulomb and Lavoisier, by working away independently in one's own home laboratory. Science today requires advanced university training, financial support, high-powered teamwork, and access to costly instruments that are well beyond the means of most individuals.

But in astronomy, non-professionals can still make important discoveries and produce useful data. This is because the sky is accessible to all and because modern technology has made small instruments of high quality available at moderate cost. It is also because the enormous number of amateurs who scrutinize the sky – hundreds of thousands of them throughout the world – almost guarantees that no new event will be missed. Most of these non-professionals are in it simply for personal enjoyment, but more than a few are involved in research programs,[†] monitoring the sky for transient phenomena such as variable stars and supernovae, or hoping to discover comets and asteroids. Some even collaborate in professional programs, one recent one being the search for exoplanets by gravitational lensing (Q. 182).

Finally, it is possible to participate in some research programs without ever observing the sky. The most famous of these projects, *SETI@home*, coordinated by the SETI program (Search for Extra-Terrestrial Intelligence) (Q. 186), uses donated time on personal computers to process signals from the giant Arecibo radio telescope in Puerto Rico. The Galaxy Zoo is another interesting project (www.galaxyzoo.org) open to members of the public, who can participate from home, via the Internet, to help classify by shape some of the one million galaxies imaged by the Sloan Digital Sky Survey.

248 You think you have made a discovery: what should you do?

You think you have discovered a new comet? A supernova? Observed intriguing behavior in a variable star? Then contact the Central Bureau of Astronomical Telegrams, the non-profit organization of the International Astronomical Union (IAU) responsible for disseminating such discoveries. You will find instructions on the Bureau's Internet website (http://www.cfa.harvard.edu/iau/cbat.html).

Obviously, before trying to alert the world, it is important to be sure that the observed phenomenon is real and actually new. If the object appears fuzzy and elongated, it could be a comet. But unless it shows an apparent displacement against the background stars over a period of an hour or more, it is more probably a distant galaxy or a parasitic reflection inside the telescope optics. Over 90% of comet "discoveries" are false alerts!

249 How does one become a professional astronomer?

Astronomers today are not the cartoon characters in white coats the public sometimes imagines, spending long, cold nights peering through a telescope, high in their mountaintop retreats. Today they are more often to be found in front of their computers, processing observations, adjusting theoretical models, writing scientific articles, and exchanging ideas with their colleagues around the world. They typically spend little time making observations – only a few nights per year. And even then, most of the time

---

[†] See, for example, the Internet website of the Center for Backyard Astrophysics
http://cbastro.org/ and of Luxorion Research Activities for Amateurs
http://www.astrosurf.com/luxorion/research1.htm.

An astronomer at work. The percentage of women in astronomy is growing. They now fill close to 15% of university faculty positions in the USA and Canada. Credit: Hertzberg Institute of Astrophysics.

they will be sitting in front of a bank of computer screens, only occasionally going to check the instrument being used on the telescope. Images and spectra are obtained via sophisticated computer programs, usually run by the observatory staff. And more and more frequently, astronomers stay comfortably in their home offices, waiting to receive their data over the Internet. This is now the practice at several remote observatories, notably in Hawaii and Chile, and is obviously necessary for space observatories. Poetry is out, technology is in!

To become an astronomer, one needs a solid knowledge of astronomy, physics, and mathematics; a doctorate is essentially a prerequisite. It is a relatively small field compared to the other sciences. In North America there are about 8 000 professional astronomers, and, since turnover is low, only about 150 or 200 job openings per year, so competition is strong.

Astronomy is not solely the reserve of astronomers. There are numerous opportunities in the field for engineers and technicians, be it in optics, electronics, software programming, or space technologies. The number of individuals working at the periphery of astronomy is at least equal to the number of actual astronomers, especially in major observatories and institutions.[†]

250 How can you find an amateur astronomy club?

It is easy to become a member of an amateur astronomy club, even if you do not own a telescope. There are clubs in most relatively large cities around the world, usually sponsoring monthly activities that include talks, workshops, and observing sessions.

A search on the Web is likely to find a club based near you, or you can consult the following umbrella organizations:

[†] Which would include the European Southern Observatory, the Gemini Observatory, ALMA, the National Optical Astronomical Observatories, the National Radio Astronomical Observatories, the Space Telescope Science Institute, and the space agencies, NASA and ESA.

- in the USA, the Astronomical Society of the Pacific (http://www.astrosociety.org) and its Nightsky Network (http://www.astrosociety.org/education/nsn/nsn.html), a nationwide coalition of astronomy clubs;
- in the UK, the British Astronomical Association (http://britastro.org/baa/);
- in Canada, the Royal Astronomical Society of Canada (http://www.rasc.ca/) and the Fédération des astronomes amateurs du Québec (http://www.faaq.org/menu.htm);
- in Australia, New Zealand, South America, and parts of Asia, IceInSpace (http://www.iceinspace.com.au) is a community website dedicated to promoting amateur astronomy in the southern hemisphere. For a list of astronomy clubs specifically in Australia and New Zealand, consult http://www.astronomy.org.au/ngn;
- finally, the website of astroplace.com (http://www.astroplace.com) provides extensive lists of astronomy clubs on all continents and even indicates which regions do not yet have clubs. If you wish to start a club, contact a few existing clubs for pointers. You may be surprised at how much support you receive.

Amateur radio astronomers have their own clubs and societies. Consult http://www.radio-astronomy.org/ for more details.

Unit conversion and basic physical and astronomical measurements

Unit conversion

1 day	86 400 s
1 sidereal day	86 164.091 s
1 arcsecond (1″)	$4.85 \cdot 10^{-6}$ radian
1 m	3.28 ft
1 cm	0.393 in
1 km	0.62 statute mile
1 nm	10^{-9} m
1 parsec (pc)	3.26 light-years $= 3.086 \cdot 10^{16}$ m
1 light-year (LY)	$9.46 \cdot 10^{15}$ m
1 astronomical unit (AU)	$1.496 \cdot 10^{11}$ m
1 Pa (pascal)	0.021 psf $= 1.45 \cdot 10^{11}$ psi
1 km/h	0.62 mph
1 kg (mass)	2.2 lbs (equivalent weight)
1 °C	5/9 °F
to convert to °F:	t °F $= 9/5(t\,°C)+32$
kelvin temperature (K)	$273.16 + °C$
1 electron volt (eV)	$1.6 \cdot 10^{-19}$ J (joule)

Physical constants

Speed of light (c)	$2.998 \cdot 10^8$ m/s
Gravitational constant (G)	$6.670 \cdot 10^{-11}$ N m^2/kg^2
Standard gravitational acceleration (g)	9.807 m/s^2
Mass of the hydrogen atom	$1.67 \cdot 10^{-27}$ kg

Astronomical parameters

Solar luminosity	$3.90 \cdot 10^{26}$ W
Solar constant	1358 W/m^2
Solar mass	$1.989 \cdot 10^{30}$ kg
Solar radius	$6.96 \cdot 10^8$ m
Earth mass	$5.976 \cdot 10^{24}$ kg
Mean radius of the Earth	6371 km
Mean distance from Earth to Moon	$3.84 \cdot 10^8$ m
Mean diameter of the Moon	3476 km
Hubble constant (H_0)	70 (km/s)/Mpc

References

[1] Alvarez, L. W. *et al.*, *Extraterrestrial cause for the cretaceous-tertiary extinction*, Science, **208** (1980), 1095.

[2] Asimov, I., *The last question*, Science Fiction Quarterly, 1956.

[3] BBC, *The Listener*, **41** (1949), 567.

[4] Becker, L. *et al.*, *Bedout: a possible end-Permian impact crater offshore north western Australia*, Science, **304**: 5676 (2004), 1469.

[5] Berger A. and Loutre, M. F., *Climate: an exceptionally long interglacial ahead?*, Science, **297** (2002), 1287.

[6] Bidle, K. D., Lee, S., Marchant, D. R., and Falkowski, P. G., *Fossil genes and microbes in the oldest ice on Earth*, PNAS, **104** (2007), 13455.

[7] Blamont, J., *Le chiffre et le songe*, Odile Jacob, 1993, 18.

[8] Borra, E. F., *Liquid mirrors*, Scientific American (Feb. 1994), 76.

[9] Brohan, P. *et al.*, *Uncertainty estimates in regional and global observed temperature changes: a new dataset from 1850*, Journal of Geophysical Research, **111** (2006).

[10] Butikov, E. I., *A dynamical picture of the oceanic tides*, American Journal of Physics, **70**: 9 (Sept. 2002), 1001, also: www.ifmo.ru/butikov/.

[11] Caldeira, K. and Kasting, J. F., *The life span of the biosphere revisited*, Nature, **360** (1992), 721.

[12] Chapman, A., *A new perceived reality: Thomas Harriot's Moon Maps*, Astronomy & Geophysics, **50**: 1 (2009).

[13] Clark, R. N., *Visual Astronomy of the Deep Sky*, Cambridge University Press, 1990.

[14] Dercourt, J., *Le temps de la Terre, une aventure scientifique*, discours à l'Académie des Sciences, 2003.

[15] Diehl, R. *et al.*, *Radioactive ^{26}Al and massive stars in the Galaxy*, Nature, **439** (2006), 45.

[16] Dohrn-van Rossum, G., *History of the Hour Clocks and Modern Temporal Orders*, University of Chicago Press, 1996.

[17] Douglas, B. C., Kearney, M. S., and Leatherman, S. P., *Sea Level Rise: History and Consequences*, Academic Press, 2001.

[18] Espenak, F. and Meeus, J., *Five millennium canon of solar eclipses: −1999 to +3000*, NASA/TP-2006, 214141, 2006.

[19] http://faculty.washington.edu/chudler/moon.html.

[20] Frebel, A. *et al.*, *Discovery of HE 1523-0901, a strongly r-process enhanced metal-poor star with detected uranium*, Astrophysical Journal, **660** (2007), L117.

[21] Glazebrook, K. *et al.*, *The Gemini Deep Deep Survey: III. The abundance of massive galaxies 3–6 billion years after the Big Bang*, Nature, **430** (2004), 181.

[22] Goldsmith, D. and Owen, T., *The Search for Life in the Universe*, University Science Books, 2002.

[23] Gribbin, J. R. and Plageman, S. H., *Jupiter Effect: The Planets as Triggers of Devastating Earthquakes*, Random House, 1976.

[24] Hallyn F., trans. *Galileo Galilei, le messager des étoiles*, Sources du Savoir, 1992.

[25] Hawking, S., *The Universe in a Nutshell*, Bantam, 2001.

[26] Imbrie, J. and Imbrie, J. Z., *Modeling the climatic response to orbital variations*, Science, **207** (1980), 943.

[27] Koestler, A., *The Sleepwalkers: A History of Man's Changing Vision of the Universe*, 1959, Penguin Books, 1986 edition.

[28] Kring, D. A. and Durda, D. D., *Trajectories and distribution of material ejected from the Chicxulub impact crater: Implications for post-impact wildfires*, Journal Geophysical Research, **107**(E6) (2002), 10.1029.

[29] Lachihze-Rey M. and Luminet, J.-P., *Figures du ciel*, Bibliothèque nationale de France, 1998, p. 286.

[30] Lu, E. T. and Love, S. G., *Gravitational tractor for towing asteroids*, Nature, **438** (2005), 177.

[31] Morelon, R., *Histoire des sciences arabes, 1: astronomie, théorique et appliquée*, Seuil, 1997.

[32] http://mwmw.gsfc.nasa.gov/mmwallsky.html.

[33] NOAA, *Restless tides*, http://co-ops.nos.noaa.gov/restles3.html.

[34] England, P., Molnar, P., and Richter, F., *John Perry's neglected critique of Kelvin's age for the Earth: a missed opportunity in geodynamics*, GSA Today, **17**: 1 (2007).

[35] Racine, R., *The historical growth of telescope aperture*, PASP, **116**: 815 (2004), 77.

[36] Labeyrie, A., *Standing wave and pellicle – a possible approach to very large space telescopes*, A&A **77**, L1–L7 (1979).

[37] Rhee, J., Song, I., and Zuckerman, B., *Warm dust in the terrestrial planet zone of a Sun-like Pleiades star: collisions between planetary embryos*, The Astrophysical Journal, **675** (2008), 77.

[38] Schaefer, B. E., *The astrophysics of suntanning*, Sky & Telescope (June 1988), 596.

[39] Schopf, J. W., *Fossil evidence of archaean life*, Philosophical Transactions of the Royal Society, Part B, **361** (2006), 869.

[40] Schrödinger, E., *What is life?*, 1944, reprinted Cambridge University Press, 2002.

[41] Trehub, A., *The Cognitive Brain*, MIT Press, 1991.

[42] Vreeland, R. H., Rosenzweig, W. D., and Powers, D. W., *Isolation of a 250 million-year-old halotolerant bacterium from a primary salt crystal*, Nature, **407** (2000), 897.

[43] Ward, P. D. and Brownlee, D., *Rare Earth: Why Complex Life is Uncommon in the Universe*, Copernicus, 2000.

[44] Wright, N., *A cosmology calculator for the World Wide Web*, PASP, **118**: 1711 (2006).

Bibliography

There are numerous books on astronomy. Below are our suggestions of several books from the elementary to more specialized and some advanced level ones. This list is far from complete, but it will help you to pick among a large choice. Do not hesitate to explore the Internet and read reviews about the books that may be of interest. Some books are several years old, but most of them are obtainable (sometimes in used version) if you search the Web carefully.

General

Croswell, K., *Magnificent Universe*, New York: Simon & Schuster, 1999.

Greene, B., *Fabric of the Cosmos*, New York: Vintage, 2005.

Hawking, S. W., *A Brief History of Time: From the Big Bang to Black Holes*, New York: Bantam Books, 1988.

Hawking, S. W., *The Universe in a Nutshell*, New York: Bantam Books, 2001.

Kirshner, R., *The Extravagant Universe*, Princeton: Princeton University Press, 2002.

Lederman, L. M., *From Quarks to the Cosmos: Tools of Discovery*, Scientific American Library, **28**, 1989.

Overbye, D., *Lonely Hearts of the Cosmos*, New York: HarperCollins, 1991.

Sagan, C., *Cosmos*, New York: Ballantine Books, 1985.

Cosmology

Gamow, G., *The Creation of the Universe*, New York: Viking Press, 1952.

Guth, A., *The Inflationary Universe*, Reading, MA: Addison-Wesley, 1997.

Livio, M., *The Accelerating Universe: Infinite Expansion, the Cosmological Constant, and the Beauty of the Cosmos*, Hoboken: John Wiley & Sons, 1999.

Rees, M., *Our Cosmic Habitat*, Princeton: Princeton University Press, 2001.

Silk, J., *A Short History of the Universe*, Scientific American Library, No. 53, 1997.

Steinhardt, P. J. and Turok, N., *Endless Universe: Beyond the Big Bang – Rewriting Cosmic History*, New York: Broadway Books, 2007.

Susskind, L., *The Cosmic Landscape*, New York: Little, Brown, 2005.

Weinberg, S., *The First Three Minutes: A Modern View of the Origin of the Universe*, New York: Basic Books, 1977.

Specific topics

Andersen, G., *The Telescope: Its History, Technology and Future*, Princeton: Princeton University Press, 2007.

Bally, J. and Reipurth, B., *The Birth of Stars and Planets*, Cambridge: Cambridge University Press, 2006.

Hill, S. and Carlowicz, M., *The Sun*, New York: Harry N. Abrams, 2006.

Hoyle, F., *The Black Cloud*, London: Roc Penguin Books, 1982.

Hubble, E., *The Realm of the Nebulae*, New Haven: Yale University Press, 1936.

Impey, C., *The Living Cosmos: Our Search for Life in the Universe*, New York: Random House, 2007.

Light, M., *Full Moon*, New York: Knopf, 2002.

Luhr, J., *The Earth*, London: Dorling Kindersley, 2007.

McCully, J. G., *Beyond the Moon: A Conversational, Common Sense Guide to Understanding the Tides*, Singapore: World Scientific, 2006.

McNab, D. and Younger, J., *The Planets*, New Haven: Yale University Press, 1999.

Schlegel, E. M., *The Restless Universe: X-ray Astronomy in the Age of Chandra and Newton*, Oxford: Oxford University Press, 2002.

Waller, W. H. and Hodge, P. W., *Galaxies and the Cosmic Frontier*, Cambridge, MA: Harvard University Press, 2003.

Ward, P. and Brownlee, D., *Rare Earth – Why Complex Life is Uncommon in the Universe*, Copernicus Books, New York: Springer, 2004.

History of astronomy

Aveni, A., *People and the Sky: Our Ancestors and the Cosmos*, New York: Thames & Hudson, 2008.

Christianson, G. E., *Edwin Hubble: Mariner of the Nebulae*, New York: Farrar, Straus Giroux, 1995.

Copernicus, N., *On the Revolutions: Nicolas Copernicus Complete Works*, Baltimore: Johns Hopkins University Press, 1992.

Ferris, T., *Coming of Age in the Milky Way*, New York: Harper Perennial, 2003.

Galileo Galilei, *Sidereus Nuncius, or the Sidereal Messenger*, Chicago: University of Chicago Press, 1989.

Hoskin, M., *The Cambridge Illustrated History of Astronomy*, Cambridge: Cambridge University Press, 1997.

Hoskin, M., *The History of Astronomy: A Very Short Introduction*, Oxford: Oxford University Press, 2003.

Hoyle, F., *Home is Where the Wind Blows: Chapters from a Cosmologist's Life*, Mill Valley, CA: University Science Books, 1994.

North, J., *Stonehenge: Neolithic Man and the Cosmos*, London: HarperCollins, 1996.

North, J., *Cosmos: An Illustrated History of Astronomy and Cosmology*, Chicago: University of Chicago Press, 2008.

Ptolemy, *Ptolemy's Almagest* (translated by G. J. Toomer), London: Duckworth, 1984.

Smith, R. W., *The Expanding Universe: Astronomy's 'Great Debate' 1900–1931*, Cambridge: Cambridge University Press, 1982.

Verschuur, G. L., *The Invisible Universe: The Story of Radioastronomy*, New York: Springer, 2006.

Amateur astronomy

Consolmagno, G. and Davis, D., *Turn Left at Orion: A Hundred Night Sky Objects to See in a Small Telescope and How to Find Them*, Cambridge: Cambridge University Press, 2000.

Dickinson, T., *NightWatch: A Practical Guide to Viewing the Universe*, Richmond Hill, Ontario: Firefly Books, 2006.

Kriege, D. and Berry, R., *The Dobsonian Telescope*, Richmond: William Bell, 1998.

Levy, D. H., *Skywatching*, San Francisco: Fog City Press, 2007.

O'Meara, S. J., *Deep-Sky Companions: The Messier Objects*, Cambridge: Cambridge University Press, 2000.

Pasachoff, J. M., Tirion, W., and Peterson, R. T., *Field Guide to Stars and Planets*, Peterson Field Guides, Boston: Houghton Mifflin, 1999.

Raymo, C., *365 Starry Nights: An Introduction to Astronomy for Every Night of the Year*, New York: Fireside, 1999.

Ridpath, I. (Ed.), *Norton's Star Atlas and Reference Handbook*, New York: Dutton, 2004.

Sinnott, R. W., *Sky & Telescope's Pocket Sky Atlas*, Cambridge, MA: Sky Publishing Corporation, 2006.

Texereau, J., *How to Make a Telescope*, Richmond: Willmann-Bell, reprinted 1984.

Advanced books

Barrow, J. and Tipler, F. J., *The Anthropic Cosmological Principle*, Oxford: Oxford University Press, 1988.

Bhatnagar, A. and Livingston, W., *Fundamentals of Solar Astronomy*, Singapore: World Scientific, 2005.

Carroll, B. W. and Ostlie, D. A., *An Introduction to Modern Astrophysics*, Reading, MA: Addison-Wesley, 1996.

Einstein, A., *The Meaning of Relativity*, Princeton: Princeton University Press, 1945.

Kragh, H. S., *Conceptions of Cosmos: From Myths to the Accelerating Universe – A History of Cosmology*, Oxford: Oxford University Press, 2007.

Kuhn, T. S., *The Copernican Revolution: Planetary Astronomy in the Development of Western Thought*, Cambridge, MA: Harvard University Press, 1957.

McFadden, L.-A., Weissman, P. R., and Johnson, T. V. (Eds.), *Encyclopedia of the Solar System*, San Diego: Academic Books, 2006.

Stacey, F. D. and Davis, P. M., *Physics of the Earth*, Cambridge: Cambridge University Press, 2008.

Internet sites

almaobservatory.org: the website of the Atacama Large Millimeter/Submillimeter Array.

astrobiology.nasa.gov: the NASA website on astrobiology.

astronomy.com: the website of the Astronomy magazine with news, updates on the Sun, the Moon, and the planets in the sky. Several articles giving good introductions to different topics of astronomy, including "How to get started in the hobby of astronomy."

astronomy.org.au: the website of the Australian Astronomical Society with sections on amateur astronomy and education.

casca.ca: the Canadian Astronomical Society website has a section on Canada's Astronomy education.

cfht.hawaii.edu: the Canada-France-Hawaii Telescope website.

chandra.harvard.edu/: the website of the Chandra X-ray Observatory.

earthquake.usgs.gov: continuous updates on earthquake activity on our planet and many other resources.

esa.int: European Space Agency website.

eso.org/public: the European Southern Observatory website.

gemini.edu: the Gemini Observatory website.

gsfc.nasa.gov/apod: a new astronomical image every day and a rich archive of astronomical images.

hkas.org.hk: website of the Hong Kong Astronomical Society with lots of information (mostly in Chinese) for amateurs.

jpl.nasa.gov: Jet Propulsion Laboratory, the NASA center that manages many of the NASA interplanetary space missions.

nasa.gov: NASA website (see also the search engine nix.nasa.gov).

nrao.edu: the website of the National Radio Astronomy Observatory.

nssdc.gsfc.nasa.gov/planetary/: a NASA website on "what's new in lunar and planetary science."

ras.org.uk: the website of the Royal Astronomical Society giving extensive information of astronomical activites in the UK. Good sections of information for everyone: schools, students, and professionals.

skyandtelescope.com/news: the website of the Sky & Telescope astronomy magazine with weekly news about astronomy.

spitzer.caltech.edu: the Spitzer Space Telescope website.

stsci.edu: the Hubble Space Telescope website; see the image gallery with thousands of images from the HST.

wikipedia.com: online encyclopedia with numerous articles on astronomy topics.

youtube.com: a very popular website with access to numerous computer simulations with excellent visualizations.

See also the websites dedicated to amateur astronomy listed in Q. 250.

Index